AF541280

VEGANISM
TREND OR MEDICINE

VEGANISM
TREND OR MEDICINE

BY

HENRY MURRAY

KNOWLEDGE BAKERS

ISBN: 9789390013470 (HB)

Published by : **Knowledge Bakers**
Pune, India
Email: knowledgebakers2019@gmail.com

Digitally Printed at : **Replika Press Pvt. Ltd.**

PREFACE

Veganism is not simply a passing trend but a lifestyle choice that encompasses ethical, environmental, and health considerations. While its rise in popularity may have been driven by social trends, the scientific evidence supporting the health benefits of a well-planned vegan diet is becoming increasingly robust. Veganism has the potential to address concerns related to animal rights, environmental sustainability, and personal well-being. However, it is crucial for individuals to approach veganism with proper knowledge and ensure their dietary choices meet all nutritional requirements. Ultimately, veganism represents a significant shift in how we view and approach our diets, offering a path towards a more compassionate, sustainable, and potentially healthier future. Veganism initially emerged as a response to ethical concerns surrounding animal cruelty and exploitation. Advocates argue that refraining from consuming animal products aligns with a compassionate lifestyle that respects the rights and welfare of sentient beings. Moreover, the environmental impact of animal agriculture, including deforestation, greenhouse gas emissions, and water pollution, has prompted many individuals to adopt veganism as a means of reducing their ecological footprint. One aspect of the debate surrounding veganism revolves around its potential as a comprehensive dietary choice that meets all nutritional requirements. Critics argue that a vegan diet may be deficient in essential nutrients, particularly vitamin B12, iron, zinc, and omega-3 fatty acids, which are primarily found in animal-based foods. However, with proper planning and awareness, a well-balanced vegan diet can provide all necessary nutrients through plant-based sources, fortified foods, and supplements. Research indicates that vegans can meet their nutritional needs and maintain good health when following a balanced and varied diet.

Scientific studies have increasingly highlighted the potential health benefits associated with a vegan diet. Plant-based diets tend to be rich in fiber, antioxidants, and phytochemicals, which can promote cardiovascular health, reduce the risk of chronic diseases, and contribute to weight management. Numerous studies have shown that vegans have lower rates of obesity, hypertension, type 2 diabetes, and certain types of cancer. Additionally, plant-based diets have been linked to improved gut health, reduced inflammation, and enhanced overall longevity. It explores the potential impact of a vegan diet on amine content. It examines how a plant-based diet may affect the intake of amines, which are organic compounds found in various foods It discusses the importance of developing nutrient-dense plant varieties to address nutritional deficiencies and improve food security and explains the various plant-based protein sources, their nutritional value, and their role in reducing the environmental impact of animal agriculture. It explores the scientific evidence supporting veganism as a viable approach to health promotion. The book aims to provide a comprehensive understanding of veganism and its implications in both personal and societal contexts.

We are grateful to all those persons as well as various books, manuals, periodicals, magazines, journals etc. that helped in the preparation of this book. In spite of the best efforts, it is possible that some errors may have occurred into the compilation and editing of the book. Further queries, constructive suggestions and criticisms for the improvement of the book are always welcomed and shall be thankfully acknowledged.

Henry Murray

Contents

Chapter 1

The Increase of Amines Content in the Intake of a Vegan Diet

Abstract

Vegetarian and vegan consumers have increased in the last years. However, the food industry is facing problems responding to this growing market, since the food safety of several plant-based products is not well established. Fruits, vegetables and fermented products, such as nut and grains milks and cheeses, may be rich sources of biogenic amines; whereas, the levels of these compounds should be considered before the inclusion on a daily diet. Biogenic amines are a class of compounds with wide physiological activities as antioxidant properties, inductors of cell division and allergic processes, and sleep, sexual and behavioral disorders. In addition to the levels of biogenic amines, the levels of some of its precursors as tryptophan, 5-hydroxytryptophan and tryptamine will be presented. The foods eaten by vegans are consumed raw, cooked, fried, fermented and mainly through homemade processing methods, which have influence on the levels of bioactive compounds from the food matrix. Exposure to processing conditions such as handling, sanitary conditions, high temperatures, preparing methods (cooking in water or oil) influencing the levels of amines, will be discussed in this chapter to enrich the knowledge on food safety associated to vegan diets.

Keywords: antioxidants, biogenic amines, histamine, tyramine, food safety

1. Introduction

Human metabolism is influenced by dietary, lifestyle, environmental and genetic factors [1]. Analysis of plasma metabolites by groups showed significant differences between meat eaters, fish eaters, vegetarians, and especially vegans [2]. Randomized clinical studies have shown that plant-based diets are associated with reduced risk of mortality and morbidity from cardiovascular disease (CVD) [3]. The association of low CVD index and vegetarian dietary patterns are the result of the constant reduction of organisms' exposure to harmful substances contained in products of animal origin (for example: saturated fat, cholesterol, heme iron, N-glycolylneuramine acid, persistent organic pollutants, polycyclic aromatic hydrocarbons, heterocyclic amines and advanced glycation end products), in addition to the increased consumption of fibers and phytochemicals present in whole plant foods. Phytochemicals in plant-food include carotenoids (α- and β-carotene, lycopene, phenolic compounds, vitamin C, tocopherols, biogenic amines, among others [4], which can act synergistically by reducing inflammation and oxidative stress, providing protection against CVD [3, 5].

Biogenic amines are aliphatic organic bases of low molecular weight and have biological activity in microorganisms, plants and animals. Polyamines (putrescine, cadaverine, spermidine and spermine) and biogenic amines (serotonin, dopamine, histamine, tyramine, among others) are called bioactive amines (BAs) and are relevant for both shelf life and final product quality, as well as for human health [6]. Some polyamines play an important role in growth and can act as antioxidants [7–9], while other amines are neuroactive or vasoactive [6]. In addition, amines are also described as being indicators of plant-food safety and some countries already limit the amount of some BAs, mainly in fermented foods [10, 11].

Although many BAs, such as histamine, tyramine and putrescine are necessary for many functions in humans, consumption of foods containing large amounts of these amines can have toxicological effects. For example, excessive consumption of histamine can induce histaminic intoxication and is mainly related to heart disease (hypotension and palpitations) and headache [12]. Tyramine is also considered harmful to the body [13]. Even though some studies show the levels of amines in plant-based food [7–9, 14], the number of studies that establish legal limits in foodstuffs is still insufficient.

While histamine and tyramine can have adverse effects, the ingestion of foods rich in some amines, such as spermidine and spermine, has been linked to longevity, both in humans [15] and in plants [16]. Diamines such as putrescine and cadaverine have been described due to their occurrence in higher levels in contaminated foods or in senescent plant tissue. Besides these polyamines, monoamines such as serotonin and dopamine have important neurological and antioxidant functions in both animals and plants. These amines can be obtained by eating some foods plant-based (banana, cauliflower, grapes - *in natura* and byproducts) and/or animal (fish and meat). However, the synthesis and *turnover* of serotonin depends on the intake of the amino acid tryptophan for the formation of serotonin in the brain, since this neutrotransmitter does not cross the blood–brain barrier [17]. In mammals, serotonin is a neurotransmitter that acts on the central nervous system, affecting appetite, sleep, anxiety and mood. In addition to being an important precursor to the formation of serotonin in the brain, tryptophan is also a precursor to melatonin, an indolamine, which has antioxidant action, besides acting in physiological processes related to the regulation of circadian rhythm, mood and sleep. This chapter provides an overview of the presence of amines in plant-based diets, the impacts of food handling and processing on these molecules, in addition to the mechanisms by which these compounds are absorbed and affect physiological functions in humans.

2. Fruit, vegetables and their bioactive amines against human diseases

Fruit and vegetables play an important role in human health, since they contain many essential nutrients and phytochemicals that are responsible for preventing or reducing the risk of various chronic diseases, including cardiovascular disease, diabetes, obesity, certain types of cancer, inflammation, heart attack, stroke and septic shock. Additionally, vegans, vegetarians and omnivores, consume fruits and vegetable by-products that contribute to significant levels of important health compounds, which have been extensively researched, in addition to being profitable raw materials and easily available to the food and pharmaceutical industries.

Cells are constantly exposed to oxidizing agents and a key point is the balance of the oxidative effect by antioxidant mechanisms. To reduce the risk of developing chronic diseases in humans and possibly delay the appearance of age-related problems, nutritional recommendations are that young and old people eat at least seven portions a day of different fruits and vegetables. For a lowest risk of total cancer, studies showed that it is important to intake of 600 g/day (7.5 portions/day), however, for coronary heart disease, stroke, cardiovascular disease and all-cause mortality the lowest risk was observed at 800 g/day (10 portions/day) [18]. Many studies show a strong and positive correlation between the content of BAs and the antioxidant potential of fruits and vegetables [7, 8].

BAs are involved in several physiological processes and can act as antioxidants [7, 8] and some studies relate polyamines to ion channel regulation, DNA methylation, histone acetylation, protein biosynthesis, RNA translation, apoptosis and regulation of the immune response [19], in addition to being considered a secondary messenger, mediating some growth factors in plants [20]. Studies indicate that the BAs present in fruits and vegetables are strong antioxidant compounds with more effective free radical scavenging properties, compared to some natural antioxidants, for example, phenolic compounds [8, 21] or well-accepted synthetic compounds, such as α-tocopherol, octyl gallate and palmitoyl-ascorbic acid [22]. In fruits and vegetables, around 22 BAs have already been described, and the most detected BAs are tyramine, putrescine, cadaverine, histamine, spermine and spermidine [10], while few studies report the detection of serotonin and dopamine [7, 8].

In addition to the ingestion of BAs by food, the body synthesizes endogenous amines by producing intestinal bacteria [19, 23], which can promote excess of these amines, facilitating the increase of diseases, mainly involving high cell proliferation. Amines such as spermidine have been described as being related to increased longevity. Spermidine - the most absorbed polyamine from the human gut - is most consumed by women and is related to the increase of survivability in humans [15], due to the capacity to restore or induce efficient autophagy [24], among others factors. However, ingestion of high levels can promote cancer development when critical immunoregulatory circuits are afflicted [25]. Studies demonstrate the lesser cytotoxic effect of spermine using an *in vitro* human intestinal cell model, compared to spermidine [26]. The cytotoxic level of spermidine ranges from 5 mM (NOAEL) to 10 mM (LOAEL), while for spermime, the levels are lower (NOAEL - 2.40 mM and LOAEL - 3.23 mM) [26]. Autophagy plays an important role in the prevention of several diseases [27], who demonstrated that spermidine may decrease the level of lipids and necrotic core formation.

In plants, putrescine is formed from ornithine or arginine by the action of ornithine decaborxylase (ODC, EC 4.1.1.17) or arginine decarboxylase (ADC, EC 4.1.1.19). The addition of aminopropyl groups from S-adenosylmethionine (SAM) to putrescine is catalyzed by S-adenosylmethionine (SAM) decarboxylase (SAMDC, EC 4.1.1.50) forming spermidine, through spermidine synthase. A new aminopropyl group is added to spermidine, forming spermine by the action of spermine synthase. In contrast, the action of 1-aminocyclopropane-1-carboxylate (ACC) synthase leads to the formation of 1-aminocyclopropane-1-carboxilic acid (ACC), a precursor to ethylene [16]. Thus, ethylene and the polyamines spermidine and spermine use the same precursor, that is, SAM. Several studies have shown that higher levels of spermidine and spermine indicate juvenility, and the Put/(Spd + Spm) > 1 ratio would indicate the formation of ethylene, related to senescence. Evidently, the levels of these amines affect post-harvest life and the action of ethylene may be an indicator of a decrease in the content of spermidine and spermine. Thus, juvenile fruits and vegetables contain higher levels of spermidine and spermine and their consumption can lead to improved health, in addition to the fact that diets with adequate levels of these polyamines can increase longevity.

3. Factors that affect the content of bioactive amines in foods

Fresh fruits and vegetables contain BAs as endogenous components and, due to uncontrolled microbial enzymatic activity, they can be accumulated during storage or even after some type of post-harvest processing [8, 28]. Amines are generally not destroyed during thermal processing, using high temperature, or during storage. Often there may be an increase in content due to the ease of extracting amines from food after cooking or in overripe fruits, due to cell wall degradation, making these molecules more available [7, 8]. These data point out the importance of detecting these compounds in different food matrices for a better understanding of the amino acid profile in fresh and processed products [8, 29].

The levels of spermidine and spermine in plant-food vary depending on the physiological stage, the cultivation method and the type of thermal processing. In bananas, during the fruit ripening process, there was an accumulation of putrescine, mainly in plantains. High levels of spermidine and spermine have been found in some genotypes of plantains (**Table 1**) [8]. Besides the differences between genotypes of the same species, the BAs contents are also modified by thermal processing in colored cauliflowers [7], which often provides increased palatability, digestibility and flavor [30] (**Table 1**). The lowest levels of spd and spm were found in cauliflower 'Grafitti', although all other genotypes analyzed ('Verdi de Maceratta', 'Cheddar' and 'Foratta') showed an increase with cooking (boiling, steaming or microwaving) (**Table 1**). This effect is due to the ease of extracting the compounds by softening the cell wall; however, this same beneficial effect may induce losses to cooking water due to the hygroscopicity of BAs and other antioxidant compounds, such as polyphenols. In colored green beans, boiling induced increased levels of BAs, and steaming maintained the lowest levels [30].

The processing temperature is an important measure to prevent or inhibit the formation of BAs in foods [31]. Heat treatment, such as cooking and pasteurization, can reduce the content of BAs in foods [31], recommending methods for reducing bioactive amines in mushrooms. For example, the pasteurized pickled and sterilized natural marinade of white button mushrooms showed substantially lower amounts of Spd compared to the unprocessed product (**Table 1**) [32]. The high temperature used in the processing contributes to the reduction of the microflora contained in the food, which is involved in the production of BAs [31], even though they are considered heat stable and are not destroyed by cooking, baking or even canning [33]. BAs are produced by mesophilic bacteria, especially at temperatures ranging from 20 to 37°C and, therefore, unprocessed and untreated mushrooms should preferably be stored under refrigeration conditions to avoid the accumulation amines, which can cause some type of toxicity to the human organism in excess [31]. Initial studies showed that intact fruit bodies of *Agaricus bisporas* stored for 48 h at 6°C did not exhibit the presence of Put and Cad, but when stored at room temperature (20°C) the levels of these amines increased significantly [34]. However, no amounts of Cad and Put were detected (only one of the three samples tested) after storage of mushrooms at room temperature [32]. These authors point out that the method of handling mushroom fruit bodies during harvest and after technological processing significantly influences the content of these amines during storage. Mechanical bruises caused by poor handling on soft mushroom tissues, during harvest and technological processing, can accelerate the activity of decarboxylating bacteria causing the synthesis and/or the accumulation of BAs [35].

Food	Range of bioactive amine							
	DOP	SER	HIS	TYR	PUT	SPD	SPM	Citation
Fermented								
Mushrooms raw (mg/kg f.w.)	—	—	nd	nd	nd–53.4	1686–2714.	nd	[32]
Mushrooms pickled, pasteurized (mg/kg f.w.)	—	—	nd	nd	462.0–716.7	470.2–1129.3	nd	[32]
Mushrooms marinade, sterilized (mg/kg f.w.)	—	—	nd	nd	38.0–141.6	247.5–264.9	nd–34.19	[32]
Mushrooms 48 h storage, 6°C (mg/kg d.w.)	—	—	—	—	nd	—	—	[34]
Mushrooms 48 h storage, 20°C (mg/kg d.w.)	—	—	—	—	368.0	—	—	[34]
Jurubeba picked (vinegar), after 1 h (mg/kg f.w.)	—	—	8.5–75.3	33.9–105.7	255.7–582.3	95.7–323.2	4.9–6.8	[14]
Jurubeba picked (vinegar) after 90 days (mg/kg f.w.)	—	—	0.1–59.6	1.6–59.8	208.3–537.1	0.4–93.1	0.2–21.4	[14]
Fermented fish product (fish sauce) (ppm)	—	—	45–1220	nd–42	2.0–243	nd–98	nd–121	[48]
Kimchi (green onion and mustard leaf) (mg/kg f.w.)	—	—	8.7–386.0	nd–181.1	nd–254.5	2.3–28.5	nd–58.6	[53]
Sauerkraut samples (mg/kg f.w.)	—	—	nd–229	nd–951	2.8–529	nd–47.0	—	[54]
Sauerkraut, after 12 months	—	—	nd	168–570	63.7–216	8.3–12.5	nd–8.6	[55]
Sauerkraut (mg/kg f.w.)	—	—	37.0	60.7	108.9	10.9	1.2	[56]
Fermented pickles (*L. plantarum* isolates) (mg/L f.w.)	—	—	nd–668	—	nd–994	—	—	[58]
Soft tofu (mg/kg f.w.)	—	—	nd–21.8	nd–7.2	15.9–42.5	nd–23.8	—	[60]
Firm tofu (mg/kg f.w.)	—	—	nd–65.3	nd–179.7	nd–306.2	nd–73.1	—	[60]

Food	Range of bioactive amine							
	DOP	SER	HIS	TYR	PUT	SPD	SPM	Citation
Household sauerkraut (mg/kg f.w.) Sterilized	—	—	nd–32.40	nd–384.0	4.0–260.0	nd–28.3	—	[54]
sauerkraut (mg/kg f.w.)	—	—	nd–26.4	26.3–345.0	18.4–359.00	nd–15.2	—	[54]
Fruits and vegetables								
Banana raw pulp (green, stg 2) (mg/100 g d.w.)	26.8–38.1	7.9–20.3	6.9–7.4	8.9–10.3	13.9–21.4	16.0–19.0	13.8–14.8	[8]
Banana raw pulp (ripe, stg 5) (mg/100 g d.w.)	26.7–34.4	7.5–13.7	6.8–8.6	9.1–9.6	17.4–27.1	15.8–17.5	13.5–14.2	[8]
Kiwi raw (µg/100 g f.w.)	—	952.0	—	—	—	—	—	[36]
Banana raw peel (mg/100 g d.w.)	32.7–305.5	8.2–74.3	11.8–118.3	9.2–14.0	16.5–25.0	15.0–23.6	15.1–31.0	[8]
Plantain raw pulp (mg/100 g d.w.)	27.7–33.7	8.9–15.5	6.8–7.2	9.2–10.0	14.7–63.6	15.7–18.6	13.5–14.4	[8]
Plantain raw peel (mg/100 g d.w.)	82.1–642.7	8.2–104.8	66.3–257.8	9.5–18.6	15.5–41.3	14.5–21.3	13.2–66.6	[8]
Banana and plantain cooked (mg/100 g d.w.)	28.0–59.5	10.3–15.1	7.4–7.8	9.2–9.6	32.4–67.1	18.0–23.9	13.8–14.5	[8]
Tomato raw (µg/100 g f.w.)	—	881.0–1244	—	—	—	—	—	[36]
Cauliflower raw (µg/100 g d.w.)	8.0–42.0	45–203.0	4.0–46.0	—	34–358.0	52.0–460.0	13.0–85.0	[7]
Cauliflower raw (µg/100 g f.w.)	—	23.0	—	—	—	—	—	[36]
Cauliflower cooked, 5 min (µg/100 g d.w.)	0.02–0.35	0.73–3.0	0.10–0.65	—	0.71–8.08	0.63–27.7	0.32–7.6	[7]
Cauliflower cooked, 10 min (pg/100 g d.w.)	0.02–0.28	0.70–4.3	0.12–0.71	—	0.78–12.0	1.20–37.4	0.55–8.2	[7]
Green beans (mg/100 g f.w.)	—	nd	nd	nd	1.48–7.78	1.27–7.36	0.20–2.01	[30]
Canned apples, 1 month after production (mg/kg f.w.)	—	—	165.86	—	—	—	—	[44]

Food	Range of bioactive amine							
	DOP	SER	HIS	TYR	PUT	SPD	SPM	Citation
Canned apples, 12 month after production (mg/kg f.w.)	—	—	432.09	—	—	—	—	[44]
Green beans cooked (mg/100 g f.w.)	—	nd	nd	nd	1.35–7.75	1.35–7.75	0.29–2.83	[30]
Brocolli raw (μg/100 g f.w.)	—	17.0	—	—	—	—	—	[36]
Spinach raw (μg/100 g f.w.)	—	19.0	—	—	—	—	—	[36]
Asparagus raw (μg/100 g f.w.)	—	55.0	—	—	—	—	—	[36]

fresh weight: f.w.; dry weight: d.w.; nd: not identificad.

Table 1.
Bioactive amine content in food.

4. Bioactive amines with neurotransmitter function present in food

As well as spermidine and spermine, other amines like monoamine serotonin, present in fruits and vegetables, have also been described for their antioxidant and anti-senescent actions in plant tissues, besides having beneficial (neurotransmitter) effects related to human health [36]. Considering the effects of serotonin in humans, diets enriched with plant-based food rich in serotonin may prevent certain diseases, such as the metabolic syndrome [36]. In humans, the ingestion of tryptophan and 5-hydroxytryptophan is essential for the formation of serotonin in the brain, since this neurotransmitter does not cross the blood – brain barrier, thus the synthesis and turnover of serotonin depends on the ingestion of these compounds through food [17]. The daily recommendation of tryptophan for adults is around 4 mg kg^{-1} body weight per day and 12 mg kg^{-1} body weight for children [17]. Thus, foods that contain higher levels of tryptophan and 5-hydroxytryptophan can help in the balance of serotonin and melatonin levels. Melatonin is produced from tryptophan and from serotonin and has also been identified in fruits and vegetables [7, 37].

Serotonin levels also vary depending on the ripening stage, as described in studies with wines grape [38] and bananas and plantains (**Table 1**) [8]. Reduced levels of amines, i.e., serotonin and dopamine in advanced stages of fruit ripening may be associated with an oxidative pathway activated during senescence, which can be considered markers of this development phase [7]. Serotonin levels can also be changed in function of the cooking (**Table 1**). An increase in the content of serotonin in banana pulps ('Pelipita') was also verified after cooking treatments in microwave with the peel (14.4%), in addition to boiling the fruit with the peel (3.8%) [8]. Cooking (i.e., boiling, steaming or microwave) induce an increase in serotonin levels in colored cauliflower, with emphasis on 'Cheddar' and 'Forata' cooked by microwave [7]. However, cooking time can be harmful (**Table 1**). Some studies on thermally processed foods indicate that prolonged exposure to high temperatures can result in substantial losses of amine compounds [39]. The frying process in *Musa* spp. fruit, for example, induced serotonin losses, and is not recommended when the objective is to ingest higher levels of this indolamine [8]. The significant decrease in serotonin has also been observed in other processes used in banana fruit (e.g., frozen fruit, ice cream and fruit nectar after pasteurization) [39].

Using fruits and vegetables with the peel for domestic processing can be a very interesting strategy to increase the intake of BAs. High amounts of catecholamines and indolamines have been identified in fruit (banana pulps) submitted to thermal processing with the peel (**Table 1**) [8]. Bananas boiled with the peel resulted in increases in serotonin content of up to 3% (cooking banana 'Pelipita') and 73% dopamine ('D'Angola'). This effect can be attributed to the migration of serotonin and dopamine from the peel to the pulp, as already verified with other bioactive compounds analyzed in fruits submitted to the cooking process [8, 40]. Fruit and vegetable peels are generally more exposed to sunlight than pulp and can protect themselves from oxidative stress caused by strong sunlight and high temperatures, producing large amounts of antioxidants.

Amines considered to be dangerous to health, mainly histamine and tyramine, do not occur only in products of animal origin or processed or fermented. People who present intolerances, such as monoamine oxidase (MAO) or diamine oxidase (DAO) deficiency, should avoid some fruits and vegetables due to the levels of these amines. Treatment with monoamine oxidase inhibitors can promote hypertension due to an increase tyramine and phenethylamine in susceptible individuals [41] Doses of 200 to 800 mg of tyramine increase the systolic blood pressure by 30 mmHg [42] and this increase can cause heart failure or brain hemorrhage [43]. The knowledge of the content of the levels of these amines in foods, whether of plant or animal origin, is important to avoid damages to health. However, few studies classify foods in relation to the levels of these amines. Despite the damages, histamine plays an essential role in promoting growth and metabolic activity of the gut and is active in nervous system [44].

Alcoholic and non-alcoholic fermented drinks, as well as long-ripened cheese, meat, fish, and some fruits and vegetables should be avoided by people sensitive to histamine and/or tyramine. Preserving plant-food in canned form can affect histamine and tyramine levels, as well as shelf life. The storage time increases the histamine content in canned apple; i.e., after twelve months of production, the samples showed almost three times more histamine compared to those stored for 30 days [44]. However, other species do not have the same result. In canned jurubeba, there was a decrease in the content of histamine and tyramine during the storage time (90 days), mainly in fruits preserved in vinegar, compared to those preserved in soy oil [14]. To cause toxicity, histamine levels vary between 10 and 100 mg/100 g in food [19] and the effect can be enhanced when there are high contents of putrescine and cadaverine in the food [14]. In addition, the cooking method may increase histamine levels, as described in bananas ('Pelipita') cooked without the peel in microwave (**Table 1**) [8].

5. Fermented food and beverages and their bioactive amines against human

Amines can be found in fermented plant-based foods [45] and, in high concentrations, they can be undesired, due to causing an unpleasant aroma, in addition to physical problems such as headaches (**Table 1**). Thus, there is a growing interest in controlling the accumulation of biogenic amines using antimicrobial agents to inhibit the proliferation of amine-producing bacteria. An efficient way to control the accumulation and/or formation of undesirable amines would be to control the fermentation process and/or the introduction of spices, which can have significant potential as antimicrobial agents. The control of the fermentation process using initial cultures ensures quality control and product safety. For example, spontaneously fermented carrot juice, a novel food product, can benefit from the development of starter cultures to avoid high numbers of Enterobacteriaceae and/or high concentrations of BAs [46]. In general, *Lactobacillus plantarum* starter-culture strains are frequently used to control vegetable fermentation processes [46, 47].

Many vegetarians consume fish products. However, fish and fermented fish products (e.g., fish sauce) contain significant levels of aminoacids and BAs, some of which are undesirable, such as histamine, with levels greater than 500 ppm (**Table 1**) [48]. Besides histamine, undesirable amines such as putrescine, cadaverine and tyramine can occur in these products. Levels of 1257 and 1178 ppm of putrescine and tyramine, respectively, have been reported in fish sauce [49]. The addition of additives in the fermentation process, such as spices, can alter fermentation conditions, possibly leading to an increase or decrease in the quantities of some endogenous compounds, such as BAs in the final product [12].

In recent years, increased shelf life and food safety through the use of bioprotective microbial cultures, in particular lactic acid bacteria (LAB) and/or their antimicrobial compounds have gained great attention. One approach is based on the growing consumer demand for probiotic non-dairy products and beverages. *Lactobacillus plantarum* and *L. delbrueckii* were detected in fermented cabbage juice and can be interesting as probiotics for vegetarian and lactose intolerant consumers.

In addition, certain strains of fermented LAB, which often belong to species of the genus *Lactobacillus*, can have health-promoting effects, for example, through immunomodulation and inhibition of pathogens [46, 50]. The food product itself could be used as a new product for traditional (dairy-based) marketed probiotics, allowing the access to a new market niche, focusing on consumers who are lactose intolerant and who do not want to purchase probiotics from products of animal origin (i.e., milk) [45].

Food companies specializing in vegetable products are looking for preservation methods that guarantee functional, sensory and nutritional quality and, at the same time, microbiological safety of products. The vegetable fermentations are gaining popularity for their rich flavors and health benefits, but due to the lack of information about the microbiological process, there are concerns about food safety [46]. Despite the controversy and the negative effects of the presence of some BAs in fermented products, studies indicate that amines present in fermented plant-based foods can be used as pharmaceutical compounds to promote cardiovascular health and longevity [51]. Histamine is well known for its pro-inflammatory effects on allergy and anaphylaxis; however, several studies have demonstrated anti-inflammatory or immunoregulatory functions of histamine. For example, histamine derived from *Lactobacillus reuteri* can suppress the inflammatory action of TNF, a proinflammatory cytokine, leading to anti-inflammatory strategies for chronic immune-mediated diseases. Probiotic species can target specific signaling pathways and immune responses, that is, strains of these bacteria can represent future therapeutic agents that can serve to suppress chronic inflammation [52]. In addition, fermented plant-based foods can be an interesting source for the detection of new bacterial strains with great potential for various probiotic and industrial applications, thus the study of microbial ecosystems during the fermentation process is interesting [45].

In comparison to dairy-based food fermentations, fermentations that use plant materials as the main substrate are little explored, except for a few examples such as kimchi (fermented with *Chinese mustard*), sauerkraut (fermented with cabbage) and pickles and can be interesting to lactose intolerants, to people allergic to milk or to vegans [45]. In kimchi samples collected on the Korean market, histamine and tyramine content above safe levels were found (**Table 1**) [53]. Sauerkraut is one of the best known and most commercialized traditional vegetable foods in Europe. In a study working with 121 commercial and household samples of sauerkraut, low levels of amines such as tryptamine, spermidine and spermine were found, however, high values of putrescine and tyramine have been identified (**Table 1**) [54]. It is noteworthy that the amount of amines present in the sauerkraut depends on the market and/or on the method of preparation of the product, with variations between the works found in the literature. In sauerkraut spontaneously fermented for one year, low concentrations of spermine and spermidine have been reported, but with high concentrations of tyramine and putrescine, as the predominant amines (**Table 1**) [55]. In commercially distributed sauerkraut samples, high levels of tyramine and putrescine were also identified, but with also high concentrations of histamine [56]. The values obtained for the BAs in the sauerkraut in the different researches, place this product as of special care due to the negative effects on the consumer, taking into account that histamine and tyramine are considered toxic amines when found in high quantities.

Another important fermented vegetable product produced in the United States and Europe is pickles, which are produced by fermenting cucumber (*Cucumis sativus* L.) with lactic acid bacteria [57]. In naturally fermented pickles (pepper, cucumber, cabbage, beans, tomatoes, Armenian cucumber and mixed vegetables) by *L. plantarum* for domestic consumption, high levels of cadaverine, putrescine and histamine were found, but not in a toxic limit [58]. However, these levels must be taken into account in people with a low level of tolerance to these amines.

Soy (*Glycine max* L.) and its derivative products are nutritional solutions for vegetarians, due to their high protein content and ease in preparing foods similar to meat and milk substitutes [59]. Tofu is one of the best known products made from soy and is an important source of minerals, proteins, among others. Firm tofu is produced by pressing and containing less water, thus they contain high levels of amines, when compared to soft tofu (**Table 1**) [60]. Tolerable levels of tyramine in foods are 100 mg/kg [61]. However, depending on the sensitivity to tyramine, some care must be taken regarding the ingested amount. Small doses of tyramine can often cause severe migraines with intracranial hemorrhaging in patients treated with classic monoamine oxidase inhibitor (MAOI). However, people who use reversible MAO-A inhibitor (RIMA) can tolerate doses between 50 and 150 mg of tyramine [41, 62].

6. Conclusions

Biogenic amines are known to occur in food, and the highest concentrations are reported especially in fermented products. Despite the association of a low disease index with vegan and vegetarian dietary patterns, these people consume some processed foods. Several processed food and by-products with high amounts of proteins and amino acids, including fermented products, can contain significant amounts of BAs, impacting the food quality and safety. The antioxidant and anti-inflammatory effects of polyamines can play an important role in preventing chronic health conditions, such as cardiovascular disease and diabetes. In contrast, cancer is associated with high levels of some polyamines, caused by a change in your homeostasis. Basic care must be taken when purchasing food, which must be handled and/or industrialized under ideal hygienic quality conditions to avoid the proliferation of undesirable bacteria and, consequently, the accumulation of BAs that cause damage to health; however, more research on practical measures to reduce the BA content is needed to ensure food safety.

Acknowledgements

The authors are grateful to the National Council for Scientific and Technological Development (CNPq, Brazil) (307571/2019-0), São Paulo Research Foundation (FAPESP) (2016/22665-2 and 2019/27227-1) and Vice President Office for Graduate Studies, Universidade Estadual Paulista (UNESP) (283).

References

[1] Scalbert A, Brennan L, Fiehn O, Hankemeier T, Kristal BS, van Ommen B, et al. Mass-spectrometry-based metabolomics: Limitations and recommendations for future progress with particular focus on nutrition research. Metabolomics. 2009;**5**(4):435-458

[2] Schmidt JA, Rinaldi S, Ferrari P, Carayol M, Achaintre D, Scalbert A, et al. Metabolic profiles of male meat eaters, fish eaters, vegetarians, and vegans from the EPIC-Oxford cohort. Am J Clin Nutr. 2015;**102**(6):1518-1526

[3] Kahleova H, Levin S, Barnard ND. Vegetarian Dietary Patterns and Cardiovascular Disease. Prog Cardiovasc Dis. 2018;**61**(1): 54-61. DOI: https://doi.org/10.1016/j.pcad.2018.05.002

[4] Carlsen, M. H, Halvorsen, B. L., Hollte, K., Bohn, S. K., Dragland, S., Sampson, L., Willey, C., Senoo, H., Umezono, Y., Sanada, C., Barikmo, I., Berhe, N., Willett, W., C., Philips, K. M., Jacobs, D. R. Jr., Blomhoff R. The total antioxidant content of more than 3100 foods, beverages, spices, herbs and supplements used worldwide. Eur food Res Technol. 2010;9(3):3-376.

[5] Benzie IFF, Wachtel-Galor S. Vegetarian diets and public health: Biomarker and redox connections. Antioxidants Redox Signal. 2010;**13**(10):1575-1591

[6] Muñoz-Esparza NC, Latorre-Moratalla ML, Comas-Basté O, Toro-Funes N, Veciana-Nogués MT, Vidal-Carou MC. Polyamines in food. **6**, Frontiers in Nutrition. 2019.

[7] Diamante MS, Borges CV, da Silva MB, Minatel IO, Corrêa CR, Gomez HAG, et al. Bioactive amines screening in four genotypes of thermally processed cauliflower. Antioxidants. 2019;**8**(8).

[8] Borges CV, Belin MAF, Amorim EP, Minatel IO, Monteiro GC, Gomez Gomez HA, et al. Bioactive amines changes during the ripening and thermal processes of bananas and plantains. Food Chem. 2019;**298**:125020. DOI: https://doi.org/10.1016/j.foodchem.2019.125020

[9] Gomez HAG, Marques MOM, Borges CV, Minatel IO, Monteiro GC, Ritschel PS, et al. Biogenic Amines and the Antioxidant Capacity of Juice and Wine from Brazilian Hybrid Grapevines. Plant Foods Hum Nutr. 2020;**75**(2):258-264

[10] Papageorgiou M, Lambropoulou D, Morrison C, Kłodzińska E, Namieśnik J, Płotka-Wasylka J. Literature update of analytical methods for biogenic amines determination in food and beverages. TrAC - Trends Anal Chem. 2018;**98**:128-142

[11] Tabanelli G. Biogenic Amines and Food Quality: Emerging Challenges and Public Health Concerns. Foods. 2020;**9**(7):7-10

[12] Zhou X, Qiu M, Zhao D, Lu F, Ding Y. Inhibitory Effects of Spices on Biogenic Amine Accumulation during Fish Sauce Fermentation. J Food Sci. 2016;**81**(4):M913-M920

[13] Ruiz-Capillas C, Herrero AM. Impact of biogenic amines on food quality and safety. Foods. 2019;**8**(2).

[14] Silva MB, Rodrigues da LFOS, Monteiro GC, Monar GRS, Gomez Gomez HA, Seabra Junior S, et al. Evaluation of biogenic amines and nitrate in raw and pickled jurubeba (Solanum paniculatum L.) fruit. J Food Sci Technol. 2019;56(6):2970-8.

[15] Kiechl S, Pechlaner R, Willeit P, Notdurfter M, Paulweber B, Willeit K, et al. Higher spermidine intake is linked to lower mortality: A prospective population-based study. Am J Clin Nutr. 2018;**108**(2):371-380

[16] Lima GPP, Vianello F. Review on the main differences between organic and conventional plant-based foods. Int J Food Sci Technol [Internet]. 2011 Jan 17 [cited 2015 Aug 6];**46**(1):1-13. http://doi.wiley.com/10.1111/j.1365-2621.2010.02436.x

[17] Palego L, Betti L, Rossi A, Giannaccini G. Tryptophan biochemistry: Structural, nutritional, metabolic, and medical aspects in humans. 2016, Journal of Amino Acids. 2016. p. 1-13.

[18] Aune D, Giovannucci E, Boffetta P, Fadnes LT, Keum NN, Norat T, et al. Fruit and vegetable intake and the risk of cardiovascular disease, total cancer and all-cause mortality-A systematic review and dose-response meta-analysis of prospective studies. Int J Epidemiol. 2017;**46**(3):1029-1056

[19] Larqué E, Sabater-Molina M, Zamora S. Biological significance of dietary polyamines. Nutrition. 2007;**23**(1):87-95 http://www.ncbi.nlm.nih.gov/pubmed/17113752

[20] Harindra Champa WA, Gill MIS, Mahajan BVC, Bedi S. Exogenous treatment of spermine to maintain quality and extend postharvest life of table grapes (Vitis vinifera L.) cv. Flame Seedless under low temperature storage. LWT - Food. Sci Technol [Internet]. 2015;**60**(1):412-419 http://dx.doi.org/10.1016/j.lwt.2014.08.044

[21] Nawaz MA, Huang Y, Bie Z, Ahmed W, Reiter RJ, Niu M, et al. Corrigendum: Melatonin: Current Status and Future Perspectives in Plant Science. Front Plant Sci. 2016;7:1-2

[22] Toro-Funes N, Bosch-Fusté J, Veciana-NoguésMT, Izquierdo-PulidoM, Vidal-Carou MC. In vitro antioxidant activity of dietary polyamines. Food Res Int. 2013;**51**(1):141-147 http://www.sciencedirect.com/science/article/pii/S096399691200508X

[23] Gomez-Gomez HA, Borges CV, Minatel IO, Luvizon AC, Lima GPP. Health Benefits of Dietary Phenolic Compounds and Biogenic Amines. In 2019. p. 3-27.

[24] Eisenberg T, Knauer H, Schauer A, Büttner S, Ruckenstuhl C, Carmona-Gutierrez D, et al. Induction of autophagy by spermidine promotes longevity. Nat Cell Biol. 2009;**11**(11):1305-1314 http://dx.doi.org/10.1038/ncb1975

[25] Gostner JM, Fuchs D. Cardioprotective effect of polyamine spermidine. Am J Clin Nutr. 2019;**109**(1):218

[26] del Rio B, Redruello B, Linares DM, Ladero V, Ruas-Madiedo P, Fernandez M, et al. Spermine and spermidine are cytotoxic towards intestinal cell cultures, but are they a health hazard at concentrations found in foods? Food Chem [Internet]. 2018;**269**(July):321-326 https://doi.org/10.1016/j.foodchem.2018.06.148

[27] Michiels CF, Kurdi A, Timmermans JP, De Meyer GRY, Martinet W. Spermidine reduces lipid accumulation and necrotic core formation in atherosclerotic plaques via induction of autophagy. Atherosclerosis. 2016;**251**:319-327

[28] Jastrzebska A, Piasta A, Kowalska S, Krzemiński M, Szłyk E. A new derivatization reagent for determination of biogenic amines in wines. J Food Compos Anal. 2016;**48**:111-119

[29] Preti R, Vieri S, Vinci G. Biogenic amine profiles and antioxidant properties of Italian red wines from different price categories. J Food Compos Anal [Internet]. 2016;**46**: 7-14. http://dx.doi.org/10.1016/j.jfca.2015.09.014

[30] Preti R, Rapa M, Vinci G. Effect of Steaming and Boiling on the Antioxidant Properties and Biogenic Amines Content in Green Bean (Phaseolus vulgaris) Varieties of Different Colours. J Food Qual. 2017;**2017**:1-8

[31] Doeun D, Davaatseren M, Chung MS. Biogenic amines in foods. Food Sci Biotechnol [Internet]. 2017;**26**(6):1463-1474 https://doi.org/10.1007/s10068-017-0239-3

[32] Jabłońska-Ryś E, Sławińska A, Stachniuk A, Stadnik J. Determination of biogenic amines in processed and unprocessed mushrooms from the Polish market. J Food Compos Anal [Internet]. 2020;**92**:103492. https://doi.org/10.1016/j.jfca.2020.103492

[33] Feddern V, Mazzuco H, Fonseca FN, De Lima GJMM. A review on biogenic amines in food and feed: Toxicological aspects, impact on health and control measures. Anim Prod Sci. 2019;**59**(4):608-618

[34] Kalač P, Křížek M. Formation of biogenic amines in four edible mushroom species stored under different conditions. Food Chem. 1997;**58**(3):233-236

[35] Dadáková E, Pelikánová T, Kalač P. Content of biogenic amines and polyamines in some species of European wild-growing edible mushrooms. Eur Food Res Technol. 2009;**230**(1):163-171

[36] Islam J, Shirakawa H, Nguyen TK, Aso H, Komai M. Simultaneous analysis of serotonin, tryptophan and tryptamine levels in common fresh fruits and vegetables in Japan using fluorescence HPLC. Food Biosci [Internet]. 2016;**13**:56-59 Available from: http://dx.doi.org/10.1016/j.fbio.2015.12.006

[37] Wang SY, Shi XC, Wang R, Wang HL, Liu F, Laborda P. Melatonin in fruit production and postharvest preservation: A review. Food Chem [Internet]. 2020;320(March):126642. Available from: https://doi.org/10.1016/j.foodchem.2020.126642

[38] Murch SJ, Hall BA, Le CH, Saxena PK. Changes in the levels of indoleamine phytochemicals during véraison and ripening of wine grapes. J Pineal Res. 2010;**49**(1):95-100

[39] Plonka J, Michalski A. The influence of processing technique on the catecholamine and indolamine contents of fruits. J Food Compos Anal. 2017;**57**:102-108

[40] Tsamo CVP, Herent MF, Tomekpe K, Happi Emaga T, Quetin-Leclercq J, Rogez H, et al. Effect of boiling on phenolic profiles determined using HPLC/ESI-LTQ-Orbitrap-MS, physico-chemical parameters of six plantain banana cultivars (Musa sp). J Food Compos Anal. 2015;**44**:158-169

[41] EFSA Panel on Biological Hazards (BIOHAZ). Scientific Opinion on risk based control of biogenic amine formation in fermented foods. EFSA J. 2011;9(10):1-93.

[42] Maintz L, Novak N. Histamine and histamine intolerance. Am J Clin Nutr. 2007;**85**(5):1185-1196

[43] Naila A, Flint S, Fletcher G, Bremer P, Meerdink G. Control of biogenic amines in food - existing and emerging approaches. J Food Sci. 2010;**75**(7):R139-R150

[44] Ziarati P, Sepehr E, Heidari S. Histamine , Nitrate , and Nitrite Content in Canned and Fresh Apple Products. Biosci Biotechnol Res Asia. 2017;14(June):827-34.

[45] Wuyts S, Van Beeck W, Allonsius CN, van den Broek MF, Lebeer S. Applications of plant-based fermented foods and their microbes. Curr Opin Biotechnol [Internet]. 2020;61:45-52. Available from: https://doi.org/10.1016/j.copbio.2019.09.023

[46] Wuyts S, Beeck W Van, Oerlemans EFM. Carrot Juice Fermentations as Man-Made Microbial Ecosystems Dominated by Lactic Acid Bacteria. Appl Environ Microbiol. 2018;**84**(12):1-16.

[47] Wouters D, Grosu-Tudor S, Zamfir M, De Vuyst L. Applicability of Lactobacillus plantarum IMDO 788 as a starter culture to control vegetable fermentations. J Sci Food Agric. 2013;**93**(13):3352-3361

[48] Tsai YH, Lin CY, Chien LT, Lee TM, Wei CI, Hwang DF. Histamine contents of fermented fish products in Taiwan and isolation of histamine-forming bacteria. Food Chem. 2006;**98**(1):64-70

[49] Zaman MZ, Abdulamir AS, Bakar FA, Selamat J, Bakar J. A review: Microbiological, physicochemical and health impact of high level of biogenic amines in fish sauce. Am J Appl Sci. 2009;**6**(6):1199-1211

[50] Marco ML, Heeney D, Binda S, Cifelli CJ, Cotter PD, Foligné B, et al. Health benefits of fermented foods: microbiota and beyond. Curr Opin Biotechnol. 2017;**44**:94-102

[51] Nilsson BO, Persson L. Beneficial effects of spermidine on cardiovascular health and longevity suggest a cell type-specific import of polyamines by cardiomyocytes. Biochem Soc Trans. 2018;**47**(1):265-272

[52] Thomas CM, Hong T, van Pijkeren JP, Hemarajata P, Trinh D V., Hu W, et al. Histamine derived from probiotic *lactobacillus reuteri* suppresses tnf via modulation of pka and erk signaling. PLoS One. 2012;7(2).

[53] Lee JH, Jin YH, Park YK, Yun SJ, Mah JH. Formation of biogenic amines in pa (green onion) kimchi and gat (mustard leaf) kimchi. Foods. 2019;**8**(3).

[54] Kalač P, Špička J, Křížek M, Steidlová Š, Pelikánová T. Concentrations of seven biogenic amines in sauerkraut. Food Chem. 1999;**67**(3):275-280

[55] Kalač P, Špička J, Křížek M, Pelikánová T. Changes in biogenic amine concentrations during sauerkraut storage. Food Chem. 2000;**69**(3):309-314

[56] Mayr CM, Schieberle P. Development of stable isotope dilution assays for the simultaneous quantitation of biogenic amines and polyamines in foods by LC-MS/MS. J Agric Food Chem. 2012;**60**(12):3026-3032

[57] Zieliski H, Surma M, Zieliska D. The Naturally Fermented Sour Pickled Cucumbers. In: Fermented Foods in Health and Disease Prevention; 2016 p. 503-16

[58] Alan Y, Topalcengiz Z, Dığrak M. Biogenic amine and fermentation metabolite production assessments of Lactobacillus plantarum isolates for naturally fermented pickles. Lwt [Internet]. 2018;98:322-8. Available from: https://doi.org/10.1016/j.lwt.2018.08.067

[59] Rizzo G, Baroni L. Soy, soy foods and their role in vegetarian diets. Vol. **10**, Nutrients. 2018.

[60] Yue CS, Ng QN, Lim AK, Lam MH, Chee KN. Biogenic amine content in various types of tofu: Occurrence,

validation and quantification. Int Food Res J. 2019;**26**(3):999-1009

[61] Shalaby AR. Significance of biogenic amines to food safety and human health. Food Res Int. 1996 Oct;**29**(7):675-690

[62] McCabe-Sellers BJ, Staggs CG, Bogle ML. Tyramine in foods and monoamine oxidase inhibitor drugs: A crossroad where medicine, nutrition, pharmacy, and food industry converge. J Food Compos Anal. 2006;**19**(SUPPL.):58-65.

Chapter 2

Vegetarian or Vegan Diet: Stimulating or at Risk to Mental Health?

Abstract

Vegetarians and vegans are more preoccupied with their health and conscious of their food habits than omnivores and often have pronounced views on killing animals for food. They are generally aware of a healthy lifestyle. Their mental attitudes, strengths and vulnerabilities may differ from meat eaters. Nowadays, health considerations would seem to play a role in the decision to become vegetarian/vegan. This chapter presents an overview of the most recent scientific literature with some emphasis on aspects of the relation between psychiatric disorders and personality characteristics in subjects with a vegetarian or vegan lifestyle compared to subjects who do not follow this lifestyle.

Keywords: vegetarian, vegan, mental health

1. Introduction

There are about 1.5 billion vegetarians worldwide, only 75 millions of whom are so by choice. The others are vegetarians out of necessity. Most of the latter will start eating meat when they can afford it [1]. In this chapter we will focus on people who are vegetarian/vegan of their own free will.

The earliest description of the vegetarian diet dates back to 800 BC and is related to Buddhism and Hinduism in ancient India, whose basic belief is in the unity of all living beings. *'Ahimsa'* or non-harming is a strong element in these religions. Today, people who favour ancient Indian lifestyles or habits such as yoga, are also in favour of plant-based diets. For example, 30% of UK yoga teachers follow a plant-based diet, which is 25 times the proportion in the general UK population [2].

In the western world, Pythagoras, the ancient Greek philosopher and mathematician formulated his ideas about reincarnation, which, as a consequence, led him to avoid eating meat. At the same time, around 600 BC a number of other religious groups (e.g. Orphism) also banned eating animal products.

Pythagoras can be considered the father of ethical vegetarianism and this is where the term *Pythagorean way of life* originates. Pythagoras had followers among influential philosophers and writers and his influence on nutrition in Europe continued until the 19th century.

In medieval Europe there had been no real popular interest in vegetarian food except among some philosophers and writers. The first Vegetarian Society in England was only founded 1847, with other countries following [3]. Interest in plant-based diets have been growing ever since.

Today, people choose plant-based diets from different motives, such as health, taste/disgust, animal welfare, environmental concerns, and weight loss [4]. One reason for choosing a plant-based diet is a healthy lifestyle [5], plant-based diets give some protection against the risks of developing somatic diseases [6]. For example, a vegetarian diet has a protective effect on the incidence of and mortality from heart diseases and cancer [7] and the risk of diabetes [8, 9].

The choice of a vegetarian diet is often related to how people see themselves and how they see others. Their diet has become part of their identity [10]. Nowadays, a growing number of people eat a plant-based diet because they are concerned about climate change [3].

It is not surprising, therefore, that psychological research on everyday eating habits and changing behaviours, lifestyle aspects and their consequences is growing [11]. A recent study found a possible relationship between perceived masculinity and diet preferences with some (weak) evidence that veganism (slightly stronger for males) leads to perceptions of decreasing masculinity when compared to omnivores [12].

People have different conceptions about food and lifestyle. Because of differences in personal and psychological characteristics there may be a difference between people who use different diets. This chapter discusses whether eating a plant-based diet influences mental state, and if so: is it different from that of people with other eating habits?

2. Lifestyle

Research has shown that a healthy lifestyle has a positive influence on both physical and mental health. A longitudinal study with German (n = 2.991) and Chinese students (n = 12.405) shows that in people with a healthier lifestyle psychological well-being is higher and mental health problems are fewer compared to groups with a less healthy lifestyle. Lifestyle factors such as lower body mass index, high frequency of physical and mental activities, non-smoking, a non-vegetarian diet, and a more regular social rhythm were positive predictors of mental health [13].

Keeping companion animals is part of a lifestyle. Most companion animals are omnivores that also eat meat. In a study with 3.673 pet owners, the number of vegetarians (6.2%) and vegans (5.8%) was higher than in the general population. Only a minor part of these owners (1.6%) fed their animals a plant-based diet. All pet owners feeding their pets a plant-based diet were vegans, with the exception of one vegetarian dog owner [14].

The country with the highest percentage of vegetarians is India, with about 40% of the Asian Indians being vegetarian. A study on this population shows that a vegetarian diet does not necessarily imply a healthy lifestyle. Nowadays, people eat more refined and processed foods, fried foods and refined carbohydrates instead of whole plant foods than before. In recent years, the number of Asian Indians who needed bariatric surgery was higher in female vegetarians than in non-vegetarian females because of a higher incidence of morbid obesity [15].

Similar results were found in an Iranian population, where a plant-based diet with wholesome plants (e.g. whole grains, fruits, vegetables, nuts, legumes, vegetable oils) was associated with fewer psychological disorders and a diet with unhealthy plant foods was associated with increased risks of obesity and depression. A plant-based diet, rich in wholesome foods was inversely associated with psychological disorders. Furthermore, unhealthy plant foods (e.g. fruit juices, sweetened beverages, refined grains, potatoes, sweets/desserts) were associated with increased risks of obesity as well as depression [16].

3. Personality characteristics

Not everybody chooses a plant-based diet. Socio-demographic factors are different for meat eaters and vegetarians/vegans. In a French cross-sectional study based on self-reports and including 93.823 participants it was shown that vegetarians' educational levels were higher and vegans' were lower than those of meat eaters and vegans had a lower educational level than meat eaters. Vegetarians were more likely to be young women, and self-employed than meat eaters. Their diets were the most balanced in terms of nutrients. Vegetarians were more strict at following the French dietary guidelines than non-vegetarians, and also had fewer nutritional deficiencies in antioxidant vitamins (such as vitamin E) compared to meat eaters. Vegans, on the other hand, showed more deficiencies in some nutrients (especially vitaminB12) compared to meat eaters [17]. Also, in a study from the US found that vegetarians and vegans who had chosen this diet for health reasons were more likely to be highly educated, female and physically active [18].

Not everyone opting for a vegetarian/vegan diet continues their plant-based diet. A remarkable finding is that conservative political ideological views are a predictor of a return to eating meat [19, 20].

Semi-vegetarians' and flexitarians' motives differ from those of vegans, lacto-ovo-vegetarians, and omnivores. Semi-vegetarians and flexitarians are more vulnerable to engaging in maladaptive eating habits than those engaging in more extreme forms of meat restriction. These findings hold for all ages and genders [21].

In addition to reasons mentioned and unknown reasons, personality characteristics can also play a role. Individual differences in personality may play a significant role in explaining individual food choices. In this context, the Five-factor model of personality [22] has often been investigated. This model describes a structure of personality characteristics containing 5 factors: neuroticism, extraversion, openness to experience, agreeableness and conscientiousness.

Neuroticism, conscientiousness and extraversion have a significant direct influence on eating habits and food choices. Vegetarians and semi-vegetarians were found to be more neurotic compared to omnivores [23].

A high degree of openness to new experiences would seem to be a powerful predictor of vegetarian food choices [20, 23, 24]. On the other hand, associations between meat consumption and openness to new experiences were also found [25]. This appears to vary depending on the type of meat consumed [26]. The associations with respect to conscientiousness and meat consumption vary, with positive and negative relationships having been reported in different studies [26, 27]. Extraversion has been linked to higher meat consumption [27]; a lower degree of extraversion has been related to a lower frequency of animal product intake [28]. Vegetarians and semi-vegetarians appear to be more neurotic and depressed than omnivores [23]. Neuroticism also influences the intake of sweets and savoury foods due to emotional and external eating [25]. Studies exploring associations between agreeableness and dietary style have shown mixed results [23, 25, 27]. In a large study (n = 13.892) a comparison was made between three eating styles: *carbohydrate-based food* (e.g., bread, pasta, snacks), *meat* (e.g. red meat, poultry), and *plant-based food and fish* (e.g. vegetables, fruits, legumes, fish).

These three dietary styles showed different associations with personality. Eating plant-based food and fish was positively associated with openness, conscientiousness, and emotional stability while meat consumption was negatively associated with openness and emotional stability, and positively associated with extraversion [24].

In a study about the relation between the dark triad of personality (Machiavellianism, narcissism and psychopathy) and diet it was found that omnivores scored higher on these traits, but that these effects weakened or disappeared when corrected for gender [29].

In a small study using another method (the implicit association test) it was found that a positive attitude to plant-based diets was related to a more emphatic sensitivity towards humans and animals and also to a positive attitude towards healthy and natural products. There was a trend suggesting that vegetarians have a higher capacity to experience compassion for others who have negative experiences as compared to omnivores. Vegetarians showed a stronger association than omnivores, and flexitarians scored somewhere in between [30].

It should also be mentioned that there are studies that failed to find significant differences between vegans/vegetarians and omnivores with respect to the Big-Five personality traits, or found different results altogether.

An example is an online survey in which it was found that vegans were less neurotic than lacto-ovo vegetarians, more open and had more compatible personality traits, were more universalistic, empathic, and ethically oriented, and had a slightly higher quality of life [31].

4. Mental health

4.1 Depression

In the above a relation has been described between neuroticism and vegetarianism. It is also known that people with high neuroticism scores are more vulnerable to developing depression [32].

Depression is a mental state which is often associated with neuroticism. As mentioned above a vegetarian lifestyle is often associated with neuroticism. Therefore, it is assumed that a relation between depression and a vegetarian diet exists. Some studies show that in western culture a vegetarian diet is associated with a higher risk of depression [33]. In a longitudinal study among 9.668 male partners of pregnant women, vegetarians (3.6% of the sample) had higher depression scores on self-reports [34].

In a Chinese study of the elderly, the use of a vegetarian diet posed a higher risk for depressive symptoms, especially in men [35]. In a systematic literature review, including 18 studies with a total of 160.257 participants, 11 out of 18 studies showed that plant-based diets were associated with poorer mental health, 3 out of 18 studies showed better mental health, 4 out of 18 studies were equivocal. The higher-quality studies showed that people avoiding meat consumption ran a higher risk of depression/anxiety and/or self-harm behaviours. Despite differences in methodology and quality of the studies, the authors conclude that ending meat consumption is not a good strategy to promote psychological health [36]. In another recent systematic review and meta-analysis, including 13 studies with 17.809 individuals, it was found that vegetarians/vegans are at a higher risk of developing depressions [37].

In Seasonal Affective Disorder –winter type (SAD), an association between vegetarianism and SAD was found.

In an SAD outpatient clinic, the number of participants following a vegetarian diet was significantly higher than in the general population, and in a large group of vegetarians from Finland (from the Finnish national FINRISK 2012 study) the number of people with SAD was higher compared to omnivores [38].

Other studies arrive to different conclusions. In a study of diabetic patients in Iran, a plant-based diet seems to protect against developing depression, anxiety and stress, and these patients were better sleepers compared to meat eaters [39]. In a study of 15-year olds in four developing countries (India, Vietnam, Peru and

Ethiopia) no association was found between a vegetarian diet and emotional symptoms [40]. In a study of endurance runners, no differences in mental health were found between vegetarian, vegans and omnivores [41].

In yet another study, the authors concluded that healthy dietary patterns do matter. The authors did not just compare plant-based diets versus diets containing meat. A healthy dietary pattern containing among other things fruits, whole grain, fish, olive oil, low-fat dairy and a low intake of animal food was compared to an unhealthy dietary pattern containing processed foods, red meats, refined grains, high-fat dairy products, sweets and a low intake of fruits and vegetables. The unhealthy diet was associated with an increased risk of depression [42]. Cultural beliefs and economic circumstances may also play a role. In a large-scale multinational cross-sectional study in four different countries it was found eating a vegetarian diet was not positively or negatively associated with mental health in the US, Germany and Russia, while in China vegetarians did run a higher risk of developing depression [43].

If a relation between food and depression exists, the question arises what ingredients improve depressed mood, and more importantly what ingredients can help prevent or recover from depression. In a literature study, LaChance and Ramsey [44] found 12 antidepressant nutrients (folate, iron, long-chain omega-3 fatty acids (EPA and DHA), magnesium, potassium, selenium, thiamine, vitamin A, vitamin B6, vitamin B12, vitamin C, and zinc) which are found in plant foods like leafy greens, peppers and cruciferous vegetables, and sea-foods like mussels and oysters. These findings could lead to a ranking system of nutrients which can be used as a treatment opportunity for people with mental health issues. Another study presents some evidence for ranking plant-based food. In adolescents, regular, daily, diet of green and yellow vegetables was associated with a lower risk of depression compared to those who never ate these vegetables or only 1–2 times a week [45]. In a large cohort study (90.380 subjects) it was found that every exclusion of a food group (not exclusively animal products) from a diet was associated with the risk of developing depressive symptoms. As more food groups (meat, poultry, fish, eggs, milk and other dairy products, vegetables, legumes and/or grains) were excluded the more the risk of depression increased [46]. In addition to this, it is worth mentioning that it is still a matter of debate whether a vegetarian/vegan diet poses a risk to brain development even if supplements are added like iron, zinc and vitamin B12 [47, 48]. Some authors argue in favour of eating whole foods, not nutrients and emphasize the need for more holistic approaches in nutrition to preserve health, animal welfare, and the planet [49]. A 12-week intervention of plant-based diet, exercise, mindfulness, lifestyle and behaviour modification showed good results in the treatment of depression and anxiety, results which after 6 months still existed in most participants. In this study, it was impossible to distinguish between the different factors of the intervention, so it is unclear what the contribution of diet was to the benefits of the treatment [50].

4.2 Anxiety

A study of first year university students in the US found that vegetarians had higher perceived stress levels compared to non-vegetarians [51].

Although there is some evidence that a plant-based diet has some negative associations with mental health, there are also some studies contradicting these results. An international online survey recruited participants via diet-related social networks. Participants were divided into three group: vegans, vegetarians and omnivores. Vegans scored lower than omnivores on mood items, and male participants scored lower on anxiety scores compared to omnivores. Stress scores were lower in vegan females only. The difference between these results and other studies may be due to of methodological differences and a possible selection bias [52]. In their meta-analysis Iguacel et al. [37] found that a vegetarian/vegan diet was related to lower anxiety scores. Because of the heterogeneous character of the studies included, the authors made some sub-group analyses. In one of these subgroup analyses of anxiety it was emerged that mainly younger participants (under 26) ran a higher risk of developing anxiety. This last was a finding from the higher quality studies only.

4.3 Eating disorders

People following a vegetarian or vegan diet are very conscious of their eating habits, and so are patients with eating disorders.

In a representative German survey (n = 2.449), participants were asked if they were vegetarian/vegan (5.4%) or omnivores and then filled out the Eating Disorder Examination Questionnaire (EDE-Q8). Vegetarians/vegans scored statistically significantly higher on the EDE-Q8 [53].

A study comparing subjects without eating disorders, with a non-clinical eating disorder and with a clinical eating disorder found that the group with the most severe eating disorders contained the highest number of vegetarians/vegans [54]. A comparison between vegetarians and omnivores yielded an association between following a vegetarian diet and orthorexia (unhealthy obsession with healthy eating), while omnivores often had more cognitive restraint as well as a higher body mass index [55].

Semi-vegetarians have been defined as vegetarians who only rarely eat meat. In a study looking for associations between diets with orthorexic tendencies and depression, semi-vegetarians with strong orthorexic tendencies show more depressive symptoms than omnivores and vegetarians. The authors speculate that semi-vegetarians with orthorexic tendencies have high or pathological health-related motives to become vegetarians and have failed to do so, which can be depressogenic because of the dissonance between their conception of good food and their actual behaviour [56].

4.4 Other disorders

In a case study describing a 47-year-old female patient with a five-year history of psychosis, a serious vitamin B12 deficiency was found. Vitamin B12 in food is present in meat, fish, dairy products, and eggs. The patient had been following a strict vegan diet for seven years, and after administration of vitamin B12 the complaints disappeared. The authors point out that professionals should be aware of veganism as a cause of vitamin B12 deficiency and hence psychiatric complaints [57]. A vitamin B12 deficiency can lead to various mental disorders, such as depression, bipolar disorders, psychosis, and dementia [58]. It was found that, compared to omnivores, vegetarians often have vitamin B12 shortage and therefore are vulnerable to developing neuropsychiatric and neurological problems [59].

A cross-sectional survey in China of young children (3–6 year) showed a relation between dietary patterns and attention-deficit/hyperactivity disorder (ADHD). Unhealthy dietary patterns with processed foods and snacks were positively associated with ADHD complaints, and the more vegetarian food patterns negatively [60]. Unhealthy diet patterns may lead to a poor biochemistry status affecting ADHD behaviour. If this is the case, mental health professionals should be aware of this, in order to improve the ADHD symptomatology. No causal relationship between diet and ADHD symptoms is shown, but food preference patterns can be a consequence of ADHD behaviour [61]. Since people suffering from ADHD symptoms regularly have nutritional and vitamin deficiencies, nutritional assessments to detect potential deficits or allergies should be performed at the start of a treatment. Scientific evidence of using diet as a treatment of autism, as sometimes claimed, is weak and poor. A vegetarian diet does not lead to nutritional threats as it includes eggs and milk [62].

5. Neurologic/neuropsychologic aspects

In a *comprehensive* review of the literature on randomized clinical trials some evidence was found that vegetarian diets prevent or delay cognitive decline in elderly adults.

Some plant foods (citrus fruit, grapes, berries, cocoa, nuts, green tea and coffee) improve specific cognitive domains, most notably frontal executive functions [63].

A different conclusion can be reached by using a different review methodology when examining the literature. In a *systematic* review of the literature on the effects of plant-based diets on body and brain, no evidence was found for the putative effects of a plant-based diet. No causal relation was seen between the use of a plant-based diet and effects on cognitive functions, mental and neurological functioning, nor for any underlying mechanism [64].

6. Conclusion

In the literature there is a growing number of papers mentioning a negative relation between mental health and a vegan/vegetarian diet; this particularly holds for a relation between vegan/vegetarian diet and depression. In most studies, it is unclear if a plant-based diet leads to depression or other mental problems, or if people with mental issues choose a vegan/vegetarian diet more often. It is possible that people who are vulnerable for depression, anxiety and stress are more concerned about their own well-being, health and the fate of the Earth, and therefore make a more conscious choice of diet.

Although many relations have been found, it is good to emphasize that a large number of these studies suffer from methodological limitations. Most are cohort studies or cross-sectional studies, in which no causal relation can established. In a recent study, data of several studies were pooled and no association was found between vegetarian diet and depression (pooled data of 10 studies) and also no association between vegetarian diet and anxiety either (pooled data of 4 studies) [65].

Another recent study shows that reviewing methodology matters. In this study, where the conclusions of different reviews about the effects of diets on depression are compared, it is shown that *narrative* reviews come to stronger conclusions than *systematic* reviews with and without meta-analyses [66]. Authors' selection bias and differences in a priori assumptions for the meta-analyses may also play a role. For more robust conclusions clinical RCT trials are needed.

Some reasons (other than ethical, religious, animal welfare or health) for the growing popularity of plant-based diets are climate change and high greenhouse gas (GHG) emissions. Meat consumption is a major contributor to global warming. Looking for plant-based alternatives as a resource of proteins instead of animal foods is claimed as an attractive alternative. This can lead to a reduction in the use of arable land, nitrogen fertilizer, water and GHG emissions and therefore can lead to improved public health [67]. But there are other factors than proteins in food that are important for physical, and mental health (as discussed in this chapter).

In a small minority of people in Europe and the US environmental motives become a more popular choice, especially under young females, for the reduction of meat consumption [68].

A possible solution besides a plant-based diet could be the use of cultured meat (in vitro from animal cells), which can address the ethical, environmental and some psychological disadvantages of conventional meat production. This is a rather new area in food production, far from large scale production or social acceptance, but it might contribute to removing many drawbacks of current meat production in the future [69].

Acknowledgements

The authors are grateful to Josie Borger for the improvement of the English.

References

[1] Leahy, E., Lyons, S., & Tol, R. S. J. (2010). *An estimate of the number of vegetarians in the world*.

[2] Mace JL, McCulloch SP. Yoga, Ahimsa and Consuming Animals: UK Yoga Teachers' Beliefs about Farmed Animals and Attitudes to Plant-Based Diets. Animals. 2020;*10*(3):480. DOI: https://doi.org/10.3390/ani10030480

[3] Leitzmann C. Vegetarian nutrition: Past, present, future. American Journal of Clinical Nutrition. 2014;*100* (SUPPL. 1). DOI: https://doi.org/10.3945/ajcn.113.071365

[4] Miki, A. J., Livingston, K. A., Karlsen, M. C., Folta, S. C., & McKeown, N. M. (2020). Using Evidence Mapping to Examine Motivations for Following Plant-Based Diets. In *Current Developments in Nutrition* (Vol. 4, Issue 3). Oxford University Press. https://doi.org/10.1093/cdn/nzaa013

[5] Dyett PA, Sabaté J, Haddad E, Rajaram S, Shavlik D. Vegan lifestyle behaviors: An exploration of congruence with health-related beliefs and assessed health indices. Appetite. 2013;*67*:119-124. DOI: https://doi.org/10.1016/j.appet.2013.03.015

[6] Appleby PN, Key TJ. The long-term health of vegetarians and vegans. Proceedings of the Nutrition Society. 2016;*75*(3):287-293. DOI: https://doi.org/10.1017/S0029665115004334

[7] Dinu M, Abbate R, Gensini GF, Casini A, Sofi F. Vegetarian, vegan diets and multiple health outcomes: A systematic review with meta-analysis of observational studies. Critical Reviews in Food Science and Nutrition. 2017;*57*(17):3640-3649. DOI: https://doi.org/10.1080/10408398.2016.1138447

[8] Chiu THT, Pan WH, Lin MN, Lin CL. Vegetarian diet, change in dietary patterns, and diabetes risk: A prospective study. Nutrition and Diabetes. 2018;*8*(1):12. DOI: https://doi.org/10.1038/s41387-018-0022-4

[9] Kahleova H, Pelikanova T. Vegetarian Diets in the Prevention and Treatment of Type 2 Diabetes. Journal of the American College of Nutrition. 2015;*34*(5):448-458. DOI: https://doi.org/10.1080/07315724.2014.976890

[10] Rosenfeld DL, Burrow AL. Vegetarian on purpose: Understanding the motivations of plant-based dieters. Appetite. 2017;*116*:456-463. DOI: https://doi.org/10.1016/j.appet.2017.05.039

[11] Rosenfeld, D. L. (2018). *The psychology of vegetarianism: Recent advances and future directions*. https://doi.org/10.1016/j.appet.2018.09.011

[12] Thomas MA. Are vegans the same as vegetarians? The effect of diet on perceptions of masculinity. Appetite. 2016;*97*:79-86. DOI: https://doi.org/10.1016/j.appet.2015.11.021

[13] Velten J, Bieda A, Scholten S, Wannemüller A, Margraf J. Lifestyle choices and mental health: A longitudinal survey with German and Chinese students. BMC Public Health. 2018;*18*(1):632. DOI: https://doi.org/10.1186/s12889-018-5526-2

[14] Dodd SAS, Cave NJ, Adolphe JL, Shoveller AK, Verbrugghe A. Plant-based (vegan) diets for pets: A survey of pet owner attitudes and feeding practices. PLoS One. 2019;*14*(1):e0210806. DOI: https://doi.org/10.1371/journal.pone.0210806

[15] Borude S. Which Is a Good Diet—Veg or Non-veg? Faith-Based Vegetarianism for Protection From Obesity—a Myth or Actuality? Obesity Surgery. 2019;*29*(4):1276-1280.

DOI: https://doi.org/10.1007/s11695-018-03658-7

[16] Zamani B, Daneshzad E, Siassi F, Guilani B, Bellissimo N, Azadbakht L. Association of plant-based dietary patterns with psychological profile and obesity in Iranian women. Clinical Nutrition. 2020;*39*(6):1799-1808. DOI: https://doi.org/10.1016/j.clnu.2019.07.019

[17] Allès B, Baudry J, Méjean C, Touvier M, Péneau S, Hercberg S, et al. Comparison of Sociodemographic and Nutritional Characteristics between Self-Reported Vegetarians, Vegans, and Meat-Eaters from the NutriNet-Santé Study. Nutrients. 2017;*9*(9):1023. DOI: https://doi.org/10.3390/nu9091023

[18] Cramer H, Kessler CS, Sundberg T, Leach MJ, Schumann D, Adams J, et al. Characteristics of Americans Choosing Vegetarian and Vegan Diets for Health Reasons. Journal of Nutrition Education and Behavior. 2017;*49*(7):561-567.e1. DOI: https://doi.org/10.1016/j.jneb.2017.04.011

[19] Hodson G, Earle M. Conservatism predicts lapses from vegetarian/vegan diets to meat consumption (through lower social justice concerns and social support). Appetite. 2018;*120*:75-81. DOI: https://doi.org/10.1016/j.appet.2017.08.027

[20] Pfeiler TM, Egloff B. Examining the "Veggie" personality: Results from a representative German sample. Appetite. 2018a;*120*:246-255. DOI: https://doi.org/10.1016/j.appet.2017.09.005

[21] Forestell CA. Flexitarian Diet and Weight Control: Healthy or Risky Eating Behavior? Frontiers in Nutrition. 2018;*5*:59. DOI: https://doi.org/10.3389/fnut.2018.00059

[22] McCrae RR. The Five-Factor Model and Its Assessment in Clinical Settings. Journal of Personality Assessment. 1991;*57*(3):399-414. DOI: https://doi.org/10.1207/s15327752jpa5703_2

[23] Forestell CA, Nezlek JB. Vegetarianism, depression, and the five factor model of personality. Ecology of Food and Nutrition. 2018;*57*(3):246-259. DOI: https://doi.org/10.1080/03670244.2018.1455675

[24] Pfeiler, T. M., & Egloff, B. (2020). *Personality and eating habits revisited: Associations between the big five, food choices, and Body Mass Index in a representative Australian sample*. https://doi.org/10.1016/j.appet.2020.104607

[25] Keller C, Siegrist M. Does personality influence eating styles and food choices? Direct and indirect effects. Appetite. 2015;*84*:128-138. DOI: https://doi.org/10.1016/j.appet.2014.10.003

[26] Pfeiler TM, Egloff B. Personality and attitudinal correlates of meat consumption: Results of two representative German samples. Appetite. 2018b;*121*:294-301. DOI: https://doi.org/10.1016/j.appet.2017.11.098

[27] Pfeiler TM, Egloff B. Personality and meat consumption: The importance of differentiating between type of meat. Appetite. 2018c;*130*:11-19. DOI: https://doi.org/10.1016/j.appet.2018.07.007

[28] Medawar E, Enzenbach C, Roehr S, Villringer A, Riedel-Heller SG, Witte AV. Less animal-based food, better weight status: Associations of the restriction of animal-based product intake with body-mass-index, depressive symptoms and personality in the general population. Nutrients. 2020;*12*(5). DOI: https://doi.org/10.3390/nu12051492

[29] Sariyska R, Markett S, Lachmann B, Montag C. What Does Our Personality Say About Our Dietary Choices? Insights on the Associations Between Dietary

Habits, Primary Emotional Systems and the Dark Triad of Personality. Frontiers in Psychology. 2019;*10*:2591. DOI: https://doi.org/10.3389/fpsyg.2019.02591

[30] Cliceri D, Spinelli S, Dinnella C, Prescott J, Monteleone E. The influence of psychological traits, beliefs and taste responsiveness on implicit attitudes toward plant- and animal-based dishes among vegetarians, flexitarians and omnivores. Food Quality and Preference. 2018;*68*:276-291. DOI: https://doi.org/10.1016/j.foodqual.2018.03.020

[31] Kessler CS, Holler S, Joy S, Dhruva A, Michalsen A, Dobos G, et al. Personality profiles, values and empathy: differences between lacto-ovo-vegetarians and vegans. Forsch. Komplementarmed. 2016;**23**:95-102. DOI: 10.1159/000445369

[32] Bienvenu OJ, Brown C, Samuels JF, Liang KY, Costa PT, Eaton WW, et al. Normal personality traits and comorbidity among phobic, panic and major depressive disorders. Psychiatry Research. 2001;*102*(1):73-85. DOI: https://doi.org/10.1016/S0165-1781(01)00228-1

[33] Michalak J, Zhang XC, Jacobi F. Vegetarian diet and mental disorders: results from a representative community survey. International Journal of Behavioral Nutrition and Physical Activity. 2012;*9*(1):67. DOI: https://doi.org/10.1186/1479-5868-9-67

[34] Hibbeln JR, Northstone K, Evans J, Golding J. Vegetarian diets and depressive symptoms among men. Journal of Affective Disorders. 2018;*225*:13-17. DOI: https://doi.org/10.1016/j.jad.2017.07.051

[35] Li, X. de, Cao, H. juan, Xie, S. yu, Li, K. chun, Tao, F. biao, Yang, L. sheng, Zhang, J. qing, & Bao, Y. song. (2019). Adhering to a vegetarian diet may create a greater risk of depressive symptoms in the elderly male Chinese population. Journal of Affective Disorders, *243*, 182-187. https://doi.org/10.1016/j.jad.2018.09.033

[36] Dobersek, U., Wy, G., Adkins, J., Altmeyer, S., Krout, K., Lavie, C. J., & Archer, E. (2020). Meat and mental health: a systematic review of meat abstention and depression, anxiety, and related phenomena. In *Critical Reviews in Food Science and Nutrition* (pp. 1-14). Taylor and Francis Inc. https://doi.org/10.1080/10408398.2020.1741505

[37] Iguacel I, Huybrechts I, Moreno LA, Michels N. Vegetarianism and veganism compared with mental health and cognitive outcomes: a systematic review and meta-analysis. Nutrition Reviews. 2020. DOI: https://doi.org/10.1093/nutrit/nuaa030

[38] Meesters ANR, Maukonen M, Partonen T, Männistö S, Gordijn MCM, Meesters Y. Is There a Relationship between Vegetarianism and Seasonal Affective Disorder? A Pilot Study. Neuropsychobiology. 2016;*74*(4). DOI: https://doi.org/10.1159/000477247

[39] Daneshzad E, Keshavarz SA, Qorbani M, Larijani B, Bellissimo N, Azadbakht L. Association of dietary acid load and plant-based diet index with sleep, stress, anxiety and depression in diabetic women. British Journal of Nutrition. 2020;*123*(8):901-912. DOI: https://doi.org/10.1017/S0007114519003179

[40] Santivañez-Romani, A., Carbajal-Vega, V., & Pereyra-Elías, R. (2018). Association between a vegetarian diet and emotional symptoms: A cross-sectional study among adolescents in four developing countries. *International Journal of Adolescent Medicine and Health*, *1*(ahead-of-print). https://doi.org/10.1515/ijamh-2018-0130

[41] Wirnitzer K, Boldt P, Lechleitner C, Wirnitzer G, Leitzmann C, Rosemann T,

et al. Health Status of Female and Male Vegetarian and Vegan Endurance Runners Compared to Omnivores—Results from the NURMI Study (Step 2). Nutrients. 2018;*11*(1):29. DOI: https://doi.org/10.3390/nu11010029

[42] Li, Y., Lv, M. R., Wei, Y. J., Sun, L., Zhang, J. X., Zhang, H. G., & Li, B. (2017). Dietary patterns and depression risk: A meta-analysis. In *Psychiatry Research* (Vol. 253, pp. 373-382). Elsevier Ireland Ltd. https://doi.org/10.1016/j.psychres.2017.04.020

[43] Lavallee K, Zhang XC, Michelak J, Schneider S, Margraf J. Vegetarian diet and mental health: Cross-sectional and longitudinal analyses in culturally diverse samples. Journal of Affective Disorders. 2019;*248*:147-154. DOI: https://doi.org/10.1016/j.jad.2019.01.035

[44] LaChance LR, Ramsey D. Antidepressant foods: An evidence-based nutrient profiling system for depression. World Journal of Psychiatry. 2018;*8*(3):97-104. DOI: https://doi.org/10.5498/wjp.v8.i3.97

[45] Tanaka M, Hashimoto K. Impact of consuming green and yellow vegetables on the depressive symptoms of junior and senior high school students in Japan. PLoS One. 2019;*14*(2):e0211323. DOI: https://doi.org/10.1371/journal.pone.0211323

[46] Matta J, Czernichow S, Kesse-Guyot E, Hoertel N, Limosin F, Goldberg M, et al. Depressive Symptoms and Vegetarian Diets: Results from the Constances Cohort. Nutrients. 2018;*10*(11):1695. DOI: https://doi.org/10.3390/nu10111695

[47] Cofnas, N. (2019). Is vegetarianism healthy for children? In *Critical Reviews in Food Science and Nutrition* (Vol. 59, Issue 13, pp. 2052-2060). Taylor and Francis Inc. https://doi.org/10.1080/10408398.2018.1437024

[48] Schürmann, S., Kersting, M., & Alexy, U. (2017). Vegetarian diets in children: a systematic review. In *European Journal of Nutrition* (Vol. 56, Issue 5, pp. 1797-1817). Dr. Dietrich Steinkopff Verlag GmbH and Co. KG. https://doi.org/10.1007/s00394-017-1416-0

[49] Fardet A, Rock E. Perspective: Reductionist nutrition research has meaning only within the framework of holistic and ethical thinking. Advances in Nutrition. 2018;*9*(6):655-670. DOI: https://doi.org/10.1093/ADVANCES/NMY044

[50] Null G, Pennesi L. Diet and lifestyle intervention on chronic moderate to severe depression and anxiety and other chronic conditions. Complementary Therapies in Clinical Practice. 2017;*29*:189-193. DOI: https://doi.org/10.1016/j.ctcp.2017.09.007

[51] Olfert MD, Barr ML, Mathews AE, Horacek TM, Riggsbee K, Zhou W, et al. Life of a vegetarian college student: Health, lifestyle, and environmental perceptions. Journal of American College Health. 2020. DOI: https://doi.org/10.1080/07448481.2020.1740231

[52] Beezhold B, Radnitz C, Rinne A, Di Matteo J. Vegans report less stress and anxiety than omnivores. Nutritional Neuroscience. 2015;*18*(7):289-296. DOI: https://doi.org/10.1179/1476830514Y.0000000164

[53] Paslakis G, Richardson C, Nöhre M, Brähler E, Holzapfel C, Hilbert A, et al. Prevalence and psychopathology of vegetarians and vegans – Results from a representative survey in Germany. Scientific Reports. 2020;*10*(1):1-10. DOI: https://doi.org/10.1038/s41598-020-63910-y

[54] Zuromski KL, Witte TK, Smith AR, Goodwin N, Bodell LP, Bartlett M, et al. Increased prevalence of vegetarianism among women with eating pathology.

Eating Behaviors. 2015;*19*:24-27. DOI: https://doi.org/10.1016/j.eatbeh.2015.06.017

[55] Brytek-Matera A. Interaction between Vegetarian Versus Omnivorous Diet and Unhealthy Eating Patterns (Orthorexia Nervosa, Cognitive Restraint) and Body Mass Index in Adults. Nutrients. 2020;*12*(3):646. DOI: https://doi.org/10.3390/nu12030646

[56] Hessler-Kaufmann JB, Meule A, Holzapfel C, Brandl B, Greetfeld M, Skurk T, et al. Orthorexic tendencies moderate the relationship between semi-vegetarianism and depressive symptoms. Eating and Weight Disorders. 2020;*1*:3. DOI: https://doi.org/10.1007/s40519-020-00901-y

[57] Bachmeyer C, Bourguiba R, Gkalea V, Papageorgiou L. Vegan Diet as a Neglected Cause of Severe Megaloblastic Anemia and Psychosis. American Journal of Medicine. 2019;*132*(12):e850-e851. DOI: https://doi.org/10.1016/j.amjmed.2019.06.025

[58] Jayaram N, Rao MG, Narasimha A, Raveendranathan D, Varambally S, Venkatasubramanian G, et al. Vitamin B $_{12}$ Levels and Psychiatric Symptomatology: A Case Series. The Journal of Neuropsychiatry and Clinical Neurosciences. 2013;*25*(2):150-152. DOI: https://doi.org/10.1176/appi.neuropsych.12060144

[59] Kapoor A, Baig M, Tunio SA, Memon AS, Karmani H. Neuropsychiatric and neurological problems among vitamin B12 Deficient young vegetarians. Neurosciences. 2017;**22**(3):228-232. https://doi.org/10.17712/nsj.2017.3.20160445

[60] Yan S, Cao H, Gu C, Ni L, Tao H, Shao T, et al. Dietary patterns are associated with attention-deficit/hyperactivity disorder (ADHD) symptoms among preschoolers in mainland China. European Journal of Clinical Nutrition. 2018;*72*(11):1517-1523. DOI: https://doi.org/10.1038/s41430-018-0131-0

[61] Lange KW. Micronutrients and Diets in the Treatment of Attention-Deficit/Hyperactivity Disorder: Chances and Pitfalls. Frontiers in Psychiatry. 2020;*11*:102. DOI: https://doi.org/10.3389/fpsyt.2020.00102

[62] Cruchet S, Lucero Y, Cornejo V. Truths, Myths and Needs of Special Diets: Attention-Deficit/Hyperactivity Disorder, Autism, Non-Celiac Gluten Sensitivity, and Vegetarianism. *Annals of*. Nutrition and Metabolism. 2016;*68*(1):43-50. DOI: https://doi.org/10.1159/000445393

[63] Rajaram S, Jones J, Lee GJ. Plant-based dietary patterns, plant foods, and age-related cognitive decline. Advances in Nutrition. 2019;*10*(4):422-436. DOI: https://doi.org/10.1093/advances/nmz081

[64] Medawar, E., Huhn, S., Villringer, A., & Veronica Witte, A. (2019). The effects of plant-based diets on the body and the brain: a systematic review. In *Translational Psychiatry* (Vol. 9, Issue 1, pp. 1-17). Nature Publishing Group. https://doi.org/10.1038/s41398-019-0552-0

[65] Askari M, Daneshzad E, Darooghegi Mofrad M, Bellissimo N, Suitor K, Azadbakht L. Vegetarian diet and the risk of depression, anxiety, and stress symptoms: a systematic review and meta-analysis of observational studies. In: *Critical Reviews in Food Science and Nutrition*. Taylor and Francis Inc; 2020. DOI: https://doi.org/10.1080/10408398.2020.1814991

[66] Thomas-Odenthal F, Molero P, van der Does W, Molendijk M. Impact of review method on the conclusions of clinical reviews: A systematic review on dietary interventions in depression as a case in point. PLoS One.

2020;*15*(9):e0238131. DOI: https://doi.org/10.1371/journal.pone.0238131

[67] Eshel G, Stainier P, Shepon A, Swaminathan A. Environmentally Optimal, Nutritionally Sound, Protein and Energy Conserving Plant Based Alternatives to U.S. Meat. Scientific Reports. 2019;*9*(1):1-11. DOI: https://doi.org/10.1038/s41598-019-46590-1

[68] Sanchez-Sabate R, Sabaté J. Consumer Attitudes Towards Environmental Concerns of Meat Consumption: A Systematic Review. International Journal of Environmental Research and Public Health. 2019;*16*(7):1220. DOI: https://doi.org/10.3390/ijerph16071220

[69] Bryant, C. J. (2020). Culture, meat, and cultured meat. In *Journal of animal science* (Vol. 98, Issue 8, pp. 1-7). NLM (Medline). https://doi.org/10.1093/jas/skaa172

Chapter 3

Veganism: A New Approach to Health

Abstract

The word vegan was given by Donald Watson in 1944 in Leicester, England, who, together with several other members of the Vegetarian Society, wanted to establish a group of vegetarians who did not consume milk or dairy products. When the proposal was rejected, Watson and like-minded people founded The Vegan Society, which advocated a complete plant-based diet, excluding meat, fish, eggs, milk and dairy products (cheese, butter) and honey. Vegans do not wear fur items, wool, bone, goat, coral, pearl or any other material of animal origin. According to surveys, vegans make up between 0.2% and 1.3% of the US population and between 0.25% and 7% of the UK population. Vegan foods contain lower levels of cholesterol and fat than the usual diet.

Keywords: veganism, health, supplements

1. Introduction

Veganism is a philosophy and lifestyle that seeks to exclude the use of animals for food or clothing and includes all other forms of diet of non-animal origin. Vegan diet is based on cereals, legumes, fruits and vegetables. Vegans do not eat meat, fish, seafood, eggs, milk, dairy products, honey threads carry things made of fur, wool, bones, leather, coral, pearls or any other materials of animal origin. Within the commitment to a vegan lifestyle, there is a group of people who eat exclusively fresh raw fruits, vegetables without heat treatment. This group of vegans is called a row food diet. Veganism differs from vegetarianism in that it is reduced entirely to a plant-based diet, while vegetarians also eat some products of animal origin, when animals are not killed when obtaining these products, e.g. eggs, honey, milk and dairy products. It seems that feminist ecology has more sympathy to movements related to animal rights, because females are exactly the most explored ones by the industry: for milk, eggs, frequent pregnancies, rape, etc., which draws more empathy in women.

The word vegan was given by Donald Watson in 1944 in Leicester, England, who, with several other members of the Vegetarian Society, wanted to establish a subgroup of vegetarians who do not consume milk or dairy products. After rejecting the proposal, Watson and associates founded. The Vegan Society which advocates a complete plant-based diet. The newly formed association agreed that the cessation of any form of animal exploitation was necessary to create a much more reasonable and humane society.

People become vegan for many reasons, including ethical care for animals and the natural environment, as well as circumstances related to health, spirituality, and religion [1]. As far as ethical principles are concerned, a vegan diet prevents the mass breeding and systematic slaughter of large numbers of animals on farms. Vegan food contains lower levels of cholesterol and fat than the usual diet. From the ecological point of view, the meat industry participates in the pollution of air, land and water, contributes to the exploitation and deforestation of forests and large land areas for the cultivation of crops intended for feeding a large number of farmed animals.

The basis of a healthy vegan diet is all vegetables, vegetable proteins, good fats and whole grains. However, vegans should pay attention to the intake of calcium, magnesium and vitamin D. Adequate intake of vitamin B12 is especially important, which is available in certain types of herbal drinks such as herbal drinks that do not contain lactose or added sugars and are especially suitable for vegans. A well-planned vegan diet can meet all the needs of the body. A poorly organized vegan diet can cause a lack of calcium, iodine, iron, vitamin B12 and vitamin D, which must then be taken from vitamin and mineral supplements [2].

According to surveys, vegans make up between 0.2% and 1.3% of the American population and between 0.25% and 7% of the general population [3].

The benefits of a vegan diet are great - it reduces the risk of cancer, cardiovascular disease, myocardial infarction, stroke, rheumatoid arthritis, high blood pressure, asthma, allergies and kidney stones.

Many studies suggest a link between cancer and diet [4]. More than half of these cancer cases are potentially preventable. Diet affects approximately 30% of all cancers in developed countries and 20% in developing countries [5].

The diet enables the assessment of the connection between the disease and the intake of certain foods in relation to the usual diet [6]. Several studies have been published that deal with the relationship between dietary factors and overall cancer risk. The vegan diet is thought to be inversely related to the overall incidence of cancer. There are studies whose results for certain cancers are not in line with the diet. This lack of clarity may result from the heterogeneity of vegetarian diets among respondents in different countries, as they may vary widely in relation to the ratio of animal and plant foods eaten, food quality, cooking methods, limitations of measures used to quantify dietary nutrition, and other factors may affect the development of cancer [7, 8].

Vegetarians and vegans generally include greater amounts of plant foods, avoid the intake of meat, and often adopt other healthy lifestyles compared to non-vegetarians [9]. Thus there is reason to suspect that vegetarian diets may protect against cancer. Factors associated with the high fiber content in vegetarian diets promote increased insulin sensitivity [10]. Plant-based diet is associated with lower circulating levels of total IGF-I and higher levels of IGFBP-I and IGFBP-2 compared with a meat-eating or even a lacto-ovo-vegetarian diet [11]. Insulin and IGF-I act as promoters for most normal and pre-neoplastic tissues. Therefore, their down- regulation may reduce cancer rates [12]. The strongest evidence linking specific foods to decrease risk of certain cancers includes the consumption of fruits and vegetables and whole grains [13]. Studies show a strong inverse relationship between dietary fiber intake and colon cancer in populations at low risk for the disease [14]. Several hypotheses have been postulated to explain this effect, 2 of which are outlined as follows. First, fiber increases the bulk of the stool, which decreases transit time and, in turn, could decrease the time colonic epithelial cells are exposed to potential fecal carcinogens. Second, bacteria in the gut can ferment fiber to short-chain fatty acids such as butyrate. These fatty acids may promote colonic cell differentiation and normal cell apoptosis [15]. It is estimated that an increase of dietary fiber to 20 g a day from average current intakes would reduce the rate of colorectal cancer by 40% [16].

The vegan diet has been found to be an effective means of preventing and treating cardiometabolic diseases. The risk of type 2 diabetes can be reduced by 50%, preventing atherosclerosis, hypercholesterolemia and hypertension [17].

2. Vegan diet: risks

Most of the data on the vegan diet is based on the adult population. Less is known about the vegan diet of newborns and children. Children have a greater need for energy and protein, which is met by vegetables, tofu, beans, whole grains, nuts and seeds. It is recommended for 10% more protein in childhood than in adult vegans. For this reason, alternatives rich in low-fiber vegan proteins, such as tofu and seitan, may be desirable, as these foods usually result in high satiety and can support adequate protein intake. Because animal foods such as meat, dairy, eggs, fish and fowl are among the best sources of protein, vegans can sometimes lack sufficient protein in their diets. The American Dietetic Association, however, notes that protein requirements can be met if a variety of plant proteins are consumed. Complementary proteins, specifically, can be very helpful in providing all the essential amino acids required by the body. Complementary proteins are made up by two incomplete proteins, such as beans and rice, that, when combined, form a complete protein. These proteins do not need to be consumed at the same meal, just during the same day [18]. Though vegan diets are often rich in omega-6 fatty acids, they can lack sufficient amounts of omega-3 due to the restriction of omega-3-rich foods, including eggs, fish and other seafood. The omega-3 fatty acids found in animal sources, which include eicosapentaenoic acid and docosahexaenoic acid, are important for cardiovascular, brain and eye health. Plant foods rich in omega-3, such as flaxseed, only contain another type of fatty acid, known as alpha-linolenic acid. Vegans can purchase soy milk and breakfast bars fortified with DHA, as well as DHA supplements derived from microalgae. Because DHA can be retroconverted to EPA, sources containing DHA are sufficient for vegan diets. Vitamin D is important for bone health, and low levels have been linked to reduced bone mass. Sunlight exposure is best source of vitamin D. Therefore, vegans who do not get regular sun exposure need to consume dietary vitamin D through either fortified foods or supplements, since vitamin D occurs naturally in very few foods. Vegan sources of vitamin D that are often fortified include soy milk, rice milk and orange juice. Lack of vitamin B-12 is one of the biggest concerns with vegan diets, especially because vegan diets are typically rich in folacin, which can mask B-12 deficiency symptoms [6]. Severe B-12 deficiencies can result in both anemia and dementia, notes Harvard Medical School. So, if you are a vegan, be sure to consume B-12 fortified foods such as soy and rice beverages and breakfast cereals, or supplements. Because the bioavailability of iron in vegan diets is lower than in traditional diets, vegans should consume 1.8 times the amount of iron consumed by nonvegans. Good sources of iron for vegans include dried beans and dark, leafy green vegetables. Zinc is another mineral whose bioavailability is lower in vegans than in nonvegans. Vegan sources of zinc include legumes, soy products, grains and nuts. Some research suggests that because plant-based diets are low in iodine, vegans who do not consume key sources of iodine, including iodized salt and sea vegetables, are more likely to be deficient in the mineral. The American Dietetic Association notes that vegans tend to fall below the recommended intake of calcium. The calcium in certain calcium-rich vegetables, such as Swiss chard and spinach, is not absorbed well, making fortified foods such as fruit juices, breakfast cereal and soy and rice milk among the best choices for vegans [2].

3. Vegan diet and micronutrient deficiencies

A vegan diet can increase the risk of micronutrient deficiencies, particularly iodine, iron, zinc, calcium, Vitamin B12, Vitamin D, Vitamin B2, Vitamin A, n-3 fatty acids (docosahexanoic acid; DHA).

3.1 Iodine

In a vegan diet, iodine needs can be met by iodized salt or supplements to sources of algae [19].

3.2 Iron

Vegan sources of iron are from tofu (soy), chickpeas, nuts, seeds and grains. Unlike iron from animal sources, called heme iron, which is easily absorbed, non-heme iron from plants has poor bioavailability and lower absorption due to high levels of phytate and polyphenols. Vegans, as well as vegetarians, require 1.8 times more iron in the diet, compared to those who eat meat. Vitamin C may increase the absorption of non-heme iron. However, many vegan sources of iron, especially soy, nuts and sesame seeds, are food allergens. For non-allergic children, iron-fortified foods, including packaged cereals, can be an additional source of iron. During later childhood, iron deficiency is the most common micronutrient deficiency, which emphasizes the importance of adequate iron intake [20].

3.3 Zinc

Vegan sources of zinc include soy and other legumes, nuts, seeds and whole grains, as well as fortified cereals. Due to the lower bioavailability of zinc in plant foods, vegans (as well as vegetarians) may need 1.5 times more zinc than those who eat meat [21].

3.4 Calcium

Calcium is a mineral important for the development of bone density. It is found mostly in milk and dairy products, which are absent in the vegan diet. Foods of plant origin rich in calcium are vegetables, legumes and cereals, leafy vegetables, sesame seeds, almonds and dried figs. If calcium intake is insufficient, a supplement in combination with vitamin D that promotes absorption should be considered [22].

3.5 Vitamin B12

Vitamin B12 is the biggest problem in the vegan diet, perhaps because it is found almost exclusively in foods of animal origin. Constant vitamin B12 supplementation or herbal drinks enriched with vitamin B12 are suggested [23].

3.6 Vitamin D

Most research indicates that vitamin D3 of animal origin is more effective than vitamin D2 of plant origin. Humans get most of their vitamin D from sun exposure [24].

3.7 Vitamin B2

Vitamin B2, or riboflavin, is necessary for the metabolism of amino acids, carbohydrates and the development of the nervous system. The main dietary sources include milk, eggs and some meat, which is not part of the vegan diet, as well as leafy greens, fortified cereals, nuts and soy [25].

3.8 Vitamin A

Vitamin A is found in fortified foods and beverages, including milk, cod liver oil, eggs, and leafy green vegetables greens rich in beta-carotene (e.g., kale, spinach). Vegan intake was also below nutritional recommendations [25].

3.9 N-3 fatty acids (DHA)

Omega-3 fatty acids require special attention in the vegan diet. The inclusion of vegan, omega-3-containing foods, such as walnuts, ground chia seeds and ground flaxseed, is advisable. Concerns have been expressed about flaxseed processing. Currently, the safe amount of ground flaxseed is not well known and caution is advised. Alternatively, supplementary sources of preformed DHA should be considered [26].

It is important to note that a diet without animal meat and products also reduces the chances of food poisoning.

4. Link between diet, microbiota and health

The majority of microorganisms in the human intestine belong to the phyla Firmicutes (which includes Clostridium, Enterococcus, Lactobacillus and Ruminococcus) and Bacteroidetes (which includes Bacteroides and Prevotella in proportions determined in part by diet) [27].

Vegan samples had a significantly lower number of microbes compared to omnivores for four bacterial taxa: Bacteroides, Bifidobacterium, *E. coli*. coli and Enterobacteriaceae. Interestingly, the vegetarian sample also showed significantly reduced bacteria and bifidobacteria. It is important to note that vegans and vegetarians did not differ significantly from each other in these four taxa, nor did they differ in Enterobacter, Enterococcus, Clostridium, Klebsiella, or Lactobacillus, compared to each other or to omnivorous specimens. The vegan diet contains more carbohydrates and fiber than omnivores, and as such, vegan samples significantly reduced stool pH than control groups. The lower pH is strongly correlated with the reduced number of *E. coli* and Enterobacteriaceae, species that are not tolerant of acidic environments. Microbiota and pH of vegetarian stools fall on continuity between vegans and omnivores. These results suggest that the composition of the human gut changes with diet along the continuum, with the vegan diet differing most from omnivores, but not necessarily and significantly from those of other vegetarians. It is possible that the disproportionately high prevalence of this beneficial bacterium in the vegan gut is attributable to a high fiber diet. The role of dietary fiber needs to be examined in greater depth, beyond its mechanical effect of increasing stool bulk and speeding transit time. Dietary fiber also influences the intestinal environment by inhibiting pathogen adhesion, altering bacterial fermentation patterns and short chain fatty acid concentrations, modifying microbiota community profiles, and lowering stool pH [28]. A vegan diet promotes an intestinal microbiota that directly reduces the risk of metabolic diseases. Studies have noticed a link between a vegan diet and protection against autoimmune diseases. For example, an analysis of the Adventist cohort found that a vegan diet, but not a vegetarian one, was associated with a lower risk of hypothyroidism [29]. Four fecal hydrolytic enzymes, associated with toxic and inflammatory products, decreased during the vegan diet. However, these changes in fecal urease, hololglycine hydrolases, β-glucuronidase and β-glucosidase disappeared within two weeks of starting a normal diet. The authors attribute this reduction in fecal enzymes not only to bacterial activity during the dietary change, but also to the high fiber content in the vegan diet, which can affect fecal weight, transit time, and bacterial metabolism. More detailed research has focused on the vegan diet and the "extreme" raw vegan diet (live food movement) as a promising treatment for rheumatoid arthritis (RA). This possibility that a vegan diet may cause a rapid change in bowel profile was supported by studies in patients with rheumatoid arthritis in which a one-month transition to a vegan diet was sufficient to significantly alter fecal microflora, as determined by stool sample gas–liquid chromatographic profiles of bacterial cellular fatty acids [30]. Thus a patient's personal taste and cultural traditions may need to dictate whether a vegan diet is the ideal choice for medical nutrition therapy [31].

5. Conclusions

A well-planned vegan diet can meet all the body's needs. A poorly organized vegan diet can cause a deficiency of calcium, iodine, omega-3 fatty acids, iron, vitamin B12 and vitamin D, which must then be taken from vitamin and mineral supplements.

References

[1] Stepaniak, Joanne (2000). Being Vegan. McGraw-Hill Contemporary. pp. 2,6,17,148-150. ISBN 978-0737303230

[2] "Position of the American Dietetic Association: vegetarian diets.". J Am Diet Assoc 109 (7): 1266-1282. July 2009. doi:10.1016/j.jada.2009.05.027. PMID 19562864. http://www.eatright.org/About/Content.aspx?id=8357

[3] "Donald Watson". Times Online. Times Newspapers Ltd.. 2005-11-16. http://www.timesonline.co.uk/tol/comment/obituaries/article754304.ece. Retrieved 2006-09-15. "The Vegan Society's 25 members swelled steadily to the 5,000 of today. There are now an estimated 250,000 vegans in Britain."

[4] Yessenia Tantamango-Bartley , Karen Jaceldo-Siegl, Jing Fan, Gary Fraser. Vegetarian diets and the incidence of cancer in a low-risk population. Cancer Epidemiol Biomarkers Prev. 2013 ;22:286-294.

[5] Slattery M, Boucher K, Caan B, Potter J, Ma K. Eating patterns and risk of colon cancer. Am J Epidemiol 1998;148:4-16

[6] Park Y, Leitzmann MF, Subar AF, Hollenbeck A, Schatzkin A. Dairy food, calcium, and risk of cancer in the NIH-AARP Diet and Health Study. Arch Intern Med 2009;169:391-401

[7] Key T, Appleby P, Spencer E, Travis R, Roddam A, Allen N. Cancer incidence in vegetarians: results from the European Prospective Investigation into Cancer and Nutrition (EPIC-Oxford). Am J Clin Nutr 2009;89:1620S–6S.

[8] Fraser GE. Vegetarian diets: what do we know of effects on chronic common diseases? Am J Clin Nutr. 2009;89(Suppl):1607S–12S.

[9] Lanou AJ, Svenson B. Reduced cancer risk in vegetarians: an analysis of recent reports. Cancer Manag Res. 2011;3:1-8

[10] Fukagawa NK, Anderson JW, Hageman G, Young VR, Minaker KL. High-carbohydrate, high-fiber diets increase peripheral insulin sensitivity in healthy young and old adults. Am J Clin Nutr. 1990;52:524-8.

[11] Allen NE, Appleby PN, Davey GK, Kaas R, Rinaldi S, Key TJ. The associations of diet with serum insulin-like growth factor I and its main binding proteins in 292 women meat-eaters, vegetarians, and vegans. Cancer Epidemiol Biomarkers Prev. 2002;11:1441-8.

[12] Shen M-R, Hsu Y-M, Hsu K-F, Chen Y-F, Tang M-J, Chou C-Y. Insulin-like growth factor 1 is a potent stimulator of cervical cancer cell invasiveness and proliferation that is modulated by αvβ3 integrin signaling. Carcinogenesis. 2006;27(5):962-71.

[13] Williams MT, Hord NG. The Role of Dietary Factors in Cancer Prevention: Beyond Fruits and Vegetables. Nutr Clin Pract. 2005;20(4):451-9.

[14] Peters U, Sinha R, Chatterjee N, et al. Dietary fibre andcolorectal adenoma in a colorectal cancer early detection programme. Lancet. 2003; 361: 1491– 1495.

[15] Key TJ, Schatzkin A, Willet WC, Allen NE, Spencer EA, Travis RC. Diet, nutrition and the prevention of cancer. *Public Health Nutr.* 2004; 7: 187– 200.

[16] Bingham S, Riboli E. Diet and cancer: the European prospectiveinvestigation into cancer and nutrition. Nature. 2004; 4: 206– 215.

[17] Kahleova H, Levin S, Barnard N. Cardio-metabolic benefits of

plant-based diets. Nutrients. 2017;9(8):848.

[18] Academy of Nutrition and Dietetics Position of the Academy of Nutrition and Dietetics: vegetarian diets. J Acad Nutr Diet. 2016;116:1970-1980

[19] Baroni L, Goggi S, Battaglino R, Bergeglieri M, Fasan I, Filippin D, et al. Vegan nutrition for mothers and children: practical tools for healthcare providers. Nutrients. 2018;11(1):5.

[20] Monsen ER, Hallberg L, Layrisse M, Hegsted DM, Cook JD, Mertz W, et al. Estimation of available dietary iron. Am J Clin Nutr. 1978;31:134-141

[21] Saunders AV, Craig WJ, Baines SK. Zinc and vegetarian diets. Med J Aust. 2013;199(4):S17–s21.

[22] Meyer R, De Koker C, Dziubak R, Skrapac AK, Godwin H, Reeve K, et al. A practical approach to vitamin and mineral supplementation in food allergic children. Clin Transl Allergy. 2015;5(11):11.

[23] Berlin H, Berlin R, Brante G. Oral treatment of pernicious anemia with high doses of vitamin B12 without intrinsic factor. Acta Med Scand. 1968;184(4):247-268

[24] Wilson LR, Tripkovic L, Hart KH, Lanham-New SA. Vitamin D deficiency as a public health issue: using vitamin D 2 or vitamin D 3 in future fortification strategies. Proc Nutr Soc. 2017;76(3):392-399.

[25] Kristensen NB, Maden ML, Hansen TH, Alin KH, Hoppe C, Fagt S, et al. Intake of macro- and micronutrients in Danish vegans. Nutr J. 2015;14:1-10

[26] Abraham K, Buhrke T, Lampen A. Bioavailability of cyanide after consumption of a single meal of foods containing high levels of cyanogenic glycosides: a crossover study in humans. Arch Toxicol. 2016;90(3):559-574

[27] Power SE, O'Toole PW, Stanton C, Ross RP, Fitzgerald GF. Intestinal microbiota, diet and health. Br. J. Nutr. 2014;111:387-402.

[28] Zimmer J, Lange B, Frick JS, Sauer H, Zimmermann K, Schwiertz A, Rusch K, Klosterhalhalfen S, Enck P. A vegan or vegetarian diet substantially alters the human colonic faecal microbiota. Eur. J. Clin. Nutr. 2012;66:53-60.

[29] Tonstad S, Nathan E, Oda K, Fraser G. Vegan diets and hypothyroidism. Nutrients. 2013;5:4642-4652.

[30] Ling WH, Hänninen O. Shifting from a conventional diet to an uncooked vegan diet reversibly alters fecal hydrolytic activities in humans. J. Nutr. 1992;122:924-930.

[31] Khazrai YM, Defeudis G, Pozzilli P. Effect of diet on type 2 diabetes mellitus: A review. Diabetes Metab. Res. Rev. 2014;30:24-33

Chapter 4

Health Effects of Plant Foods and the Possibility of Reducing Health Risk

Abstract

Many nutritional studies point to an inverse relationship between diet with predominant consumption of plant foods and the incidence of diseases of civilization. The health benefits of plant foods come from a sufficient intake of protective nutrients, which are key food commodities of the vegetarian diet. These include vegetables, fruits, whole grains, legumes and oilseeds, including various types of nuts. The nutritional and non-nutritional components of plant foods reduce the risk of chronic diseases by various mechanisms, so a well-planned vegetarian diet is nutritionally adequate, fully healthy and able to provide health benefits in the prevention of many diseases. The data we find agree that the benefits far outweigh the potential risks.

Keywords: antioxidants, oxidants, intestinal microbiota, benefits and risks of alternative eating

1. Introduction

Despite the fact that vegetarianism has been known for several millennia, this diet still has many of its adherents as well as despisers. We have been dealing with the issue of health effects of plant foods since 2000 in partial epidemiological studies, so we decided to summarize our findings on how vegetarianism affects the population of Slovakia.

Plants are the basic components of the food chain, in which they provide all the essential mineral and organic nutrients for humans either directly or through animal production. A comprehensive, varied diet ensures nutritional quality (intake of all nutrients) and, despite the concentrations of nutrients in the consumed food mixture corresponding to the recommended nutritional doses, also the nutritional quantity. Both of these criteria for optimal diet are essential for the development and maintenance of human health [1]. Vitamin B12, vitamin D and n-3 fatty acids are lacking in plant food sources. In particular, the content of methionine and lysine (also other essential amino acids), iodine and carnitine is significantly reduced compared to animal food [2, 3]. The absorption of iron, calcium, zinc, copper, manganese and selenium can be inhibited by plant food components [4, 5]. These facts may be associated with many health risks in the population with exclusive or dominant consumption of plant foods [6, 7].

To prevent these risks, it is necessary to consume food or pharmaceutical supplements to compensate for deficient nutrients.

Saturated fats (animal source) cause hypercholesterolemia, polyunsaturated fats (vegetable oils) can lower blood cholesterol levels. Monounsaturated fatty acids (oleic acid - olive oil, rapeseed oil, sesame oil, hazelnuts, almonds) also have a hypolipidemic effect and, in addition, have a saving (non-reducing) effect on HDL-cholesterol [8]. Compared to nonvegetarians, vegetarians in the Oxford vegetarian study had a 24% reduction in mortality from ischemic heart disease (DRR/death rate ratio/ = 0.76) [9]. When the meat-consuming group was divided into regular consumers of red meat (at least once a week) and semi-vegetarians (consumption of fish and white meat less than once a week), the DRR for semi-vegetarians was 0.78 and for vegetarians 0.66. Consumption of cheese, eggs, total fat, and cholesterol has been associated with mortality from ischemic heart disease [10]. Compared to subjects who ate relatively few of the commodities and nutrients listed, the DRR in those with the highest consumption of cheese was 2.47, eggs 2.68, animal fat 3.29, and cholesterol 3.53. Analysis of 5 prospective studies also expressed mortality from cerebrovascular disease - DRR in vegetarians 0.93 [11]. The results of a study of more than 34,000 Adventists on Day 7 in California showed a significant association between beef consumption and fatal coronary heart disease. In men eating beef more than 3 times a week, the relative risk (RR) was 2.31 compared to vegetarians. Furthermore, this study found a significant protective association between nut consumption and fatal and non-fatal ischemic heart disease for both sexes (RR = 0.5 for subjects consuming nuts more than 5 times per week compared to less than 1 time/week) and a reduced risk of ischemic heart disease in subjects who prefer wholemeal bread versus white [12].

Fiber consumption can also reduce the risk of cardiovascular disease [13]. An analysis of 10 prospective studies in the US and Europe of more than 336,000 subjects showed that an increase in fiber consumption every 10 g/day was associated with a 14% reduction (RR = 0.86) in all coronary cases and a 27% reduction (RR = 0.73) in risk coronary death [14]. Other plant food components (saponins in legumes, plant proteins, antioxidant nutrients - vitamin C, vitamin E, β-carotene, selenium, polyphenols and flavonoids) are added to the beneficial effect of fiber [15]. By evaluating current knowledge about the protective effects of plant food ingredients, scientists agree on the need to consume a variety of plant foods [7]. Vegetarian studies have confirmed that subjects with low or no consumption of animal fats and with a dominant consumption of plant foods have low values of atherosclerosis risk factors and higher values of parameters with antisclerotic properties compared to the general population [14].

Cardiovascular disease can be positively influenced by the consumption of plant proteins. Vegetable proteins versus hen egg reference protein or other animal proteins have a reduced content of some essential amino acids and an increased content of some non-essential amino acids [15]. Experimental studies have shown that cholesterol-free diets containing milk protein, casein or other animal proteins, induced an increase in plasma concentrations of total and LDL-cholesterol, while plant proteins did not have such an adverse effect [16]. By selectively increasing amino acid intake, hypercholesterolemia was found to be primarily due to essential amino acids. Higher intake of lysine and methionine (from animal proteins) adversely affects the metabolism of phospholipids in the liver. Higher amounts of methyl groups from methionine may lead to increased secretion of apoB lipoproteins. These data suggest that lower methionine and lysine intake in subjects with exclusive or predominant consumption of plant proteins represents a protective effect against cardiovascular risk. In the general population, the ratio of animal to vegetable protein consumption is about 60:40, more preferably 50:50.

2. Characteristics of the monitored group of probands with different eating habits

Our research included a randomly selected adult population from Bratislava and its surroundings. A database of volunteer addresses from previous solved research projects monitoring the health and nutritional status of the population was used. The number of probands in both research groups (non-vegetarian group/general population /and alternative diet group with predominance of plant food consumption/vegetarians)) as well as other characteristics of the groups are given in **Table 1**. Examinations were performed in the spring. The intake of vitamins, minerals and trace substances was only in its natural form (no supplementation). Another limitation of random selection was the requirement that probands be subjectively healthy, i. zn. without cardiovascular disease, cancer, diabetes, kidney disease, digestive tract and thyroid disease.

Monitoring of health and nutritional status was performed by somatometric, dietary, biochemical and microbiological examination.

2.1 Somatometric examination

Somatometric examination was performed by measuring body height and weight. The body mass index (BMI = weight/height^2) was calculated to assess overweight and obesity. A calipometer, Omron and InBody 230 were used for further somatometric measurements the type of obesity was determined after recalculation. We used Omron to measure bioimpedance in the upper half of the body, but the results can be influenced by the type of obesity and data entry. InBody 230 was performed by bioimpedance measurement using DSM-BIA (segment multi-frequency bioelectric impedance) technology, ie measurement of the upper and lower body, so the results are not affected by the type of obesity.

At first glance, most vegetarians are recognizable. They have a slim figure, pale skin and are anemic. These facts were also confirmed in our study (**Table 2**). Subcutaneous fat on the abdominal algae measured with a caliper, as well as visceral fat, have been significantly reduced by vegetarians compared to non-vegetarians. Also, the percentage of fat measured using the InBody 230 and Omron BF-306, both instruments using the bioimpedance method, confirmed a significantly reduced percentage of fat in vegetarians.

	Vegetarians	**Non-vegetarians**
n	920	958
age range (y)	20–60	20–60
average age (y)	40,41 ± 0,96	40,50 ± 1,13
BMI (kg/m2)	21,46 ± 0,35***	25,60 ± 0,47
time vegetarianism (y)	18,61 ± 0,63	
smokers	0	0
Systolic blood pressure (mmHg)	120,57 ± 1,44***	143,59 ± 1,70
Diastolic blood pressure (mmHg)	72,46 ± 0,95***	88,27 ± 0,94
Pulse rate (1 minute)	70,57 ± 1,33	77,01 ± 1,25

Table 1.
Characteristics of the monitored group.

	Vegetarians	Non-vegetarians
Belly (mm) on caliper	24,40 ± 1,13***	30,16 ± 1,01
Visceral fat (cm2)	70,06 ± 4,94***	100,56 ± 4,13
% fat on InBody230	23,16 ± 0,69*	25,93 ± 0,88
% fat on Omron BF-306	27,00 ± 0,67*	28,27 ± 0,86

Table 2.
Subcutaneous and visceral fat.

2.2 Dietary examination

The nutritional regime was determined on the basis of a questionnaire on the frequency of use of selected commodities. The questionnaire focused on the amount and frequency of consumption of 144 individual foods, food groups and recipes. Groups and recipes include soups, soups, sauces, pickled vegetables, fruits and jams. Probands responded to the consumption of 4 categories: almost never, x times a month, x times a week, x times a day. The data after dispensing into individual foods as well as after the exact expression of the amounts of consumption and conversion to daily intake were processed using the Nutrition program, which is a food data bank of the Food Research Institute in Bratislava. The calculation revealed the loss of vitamins during technological food processing. The output of the evaluation of the nutritional regime was the average daily energy intake and selected nutrients.

2.3 Biochemical examination

Collection of biological material - blood, urine and stool.

Blood was collected in the morning on an empty stomach after standard food intake in the previous days to examine the monitored parameters. Blood collection for plasma was performed in commercial syringes with EDTA (ethylene diamine tetraacetate), which in addition to its anticoagulant properties is also an inhibitor of free radical reactions. Morning urine and feces were also collected and processed and stored at −80° C on the day of collection. Average sample 24-h. urine volume for the determination of iodine probandi brought in the morning on the day of examination.

Total cholesterol, HDL-cholesterol, triacylglycerols, glucose and iron were determined by standard laboratory methods with a Vitros 250 automated analyzer (Johnson & Johnson, USA). The LDL-cholesterol content was determined by calculation from the Friedewald formula: LDL-cholesterol = total cholesterol - triacylglycerols/2,2 - HDL-cholesterol (6). The condition of the calculation was the value of triacylglycerols <4.5 mmol/l. The atherogenic index was calculated from the ratio of total and HDL-cholesterol. Vitamin C, E, A and β-carotene, malondialdehyde were measured by HPLC [17–20]. Protein carbonyls, conjugated dienes were determined spectrophotometrically [21]. The alkaline comet assay was used to detect DNA breaks, oxidized purines and oxidized pyrimidines in isolated lymphocytes [22–24]. Determination of 25-hydroxyvitamin D and other hydroxylated metabolites of vitamin D (both its forms - ergocalciferol D2 and cholecalciferol D3) in serum was performed by the classical equilibrium RIA method. Plasma N-carboxymethyllysine and plasma fluorescent AGEs were determined by competitive ELISA [25]. The iodine value was determined in the 24-h samples. Urine by modification of the Sandell-Kolthoff reaction [26].

2.4 Microbiological examination

The qualitative and quantitative representation of the microflora was evaluated by classical microbiological methods by culturing on selective diagnostic nutrient media according to Mitsuok et al. [27]. In probands for which we did not notice differences in the qualitative representation by culture methods, we determined bacterial profiles using molecular biology methods. The presence of potentially mutagenic substances was determined by Ames plate incorporation assay using S. typhimurium TA98 (shift mutations) and TA100 (point mutations) bacterial cultures. Sterol analysis by gas chromatography was performed using a GC17-QP5000, with a split/splitless injector (Shimadzu, Kyoto, Japan).

2.5 Statistical processing of results

Commercial programs - Excel 2000 and Statgraphics for Windows, version 1.4 and PASW Statistics 18 were used for computer processing of the obtained data. The comparison was supplemented by determining the percentage occurrence of risk values of monitored parameters as well as the occurrence of protective values of antioxidant vitamins in each evaluated group. This partially eliminated the fact that malnutrition probands could also be included in the traditional diet (this is a current sample of our population, whose malnutrition ultimately reflects a high incidence of morbidity and mortality from the two main diseases - cardiovascular and cancer), while in vegetarians, incorrect vegetarianism is less common (the risks are given by the one-sidedness of the diets resulting from the type of alternative diet).

3. Health risks of dominant consumption of vegetable food

3.1 Vitamin B12

Vitamin B12 is absent from plant foods; bacteria in the lower part of the small intestine are its only source in subjects with exclusive consumption of plant foods, therefore vitamin B12 deficiency is one of the risk factors for alternative diets [28]. Vitamin deficiency can have many adverse health consequences: folate "flap" in the methylation cycle, deterioration of DNA biosynthesis, pernicious anemia, increased atherogenic homocysteine in the blood, and neural tube defects [29]. Consumption of dairy products and eggs in lacto-ovo-vegetarians and, in addition, intake of white meat in semi-vegetarians provides a better ability to meet the body's vitamin B12 needs [28, 30]. In the monitored groups of volunteers, we found significantly reduced concentrations of vitamin B12 in the group of vegans (VV) and lacto-ovo-vegetarians (V-LO) and insignificantly lower serum concentration in semivegetarians (VS) compared to the traditional diet of the general population (NV). Deficiency values occurred in 67% V-V, 32% V-LO, 7% V-S, but no non-vegetarian. From a global perspective, vitamin B12 deficiency prevention requires monitoring of serum vitamin B12 levels and strict vitamin B12 fortified food or vitamin B12 supplements, especially in strict vegetarians but also in V-LO. One of the many functions of vitamin B12 is its involvement in the metabolism of homocysteine (HCy), which has atherogenic properties. HCy is a sulfur amino acid that is metabolized in two ways by B-group vitamins - remethylation (requires vitamin B9 and B12), which converts HCy back to methionine, and transsulfurization (requires vitamin B6), which converts HCy to cysteine and taurine [31]. The first of the pathways dominates with lower methionine intake, which occurs in V-LO and V-V diets, because plant proteins contain less of this amino acid. The results shows a higher incidence of mild hyperhomocysteinemia in lacto-ovo-vegetarians and vegans, in which remethylation predominates, but in vitamin B12 deficiency the remethylation cycle is inhibited and HCy is not degraded to methionine.

3.2 Vitamin D

Subjects with limited animal food intake may be at greater risk of vitamin D deficiency compared to non-vegetarians because the food that provides the highest amounts of vitamin D is of all animal origin [32]. The authors of Crowe et al. [33] reported that plasma concentrations of 25 (OH) vitamin D reflect the rate of elimination of consumption of animal products. Meat consumers had the highest average vitamin D intake (3.1 μg / day) and a mean plasma concentration of 25 (OH) D (77.0 nmol / l), vegans had the lowest average intake and plasma concentration (0.7 μg /day) and 55.8 nmol/l). Our results show that under conditions of the same and low intensity of sunlight (spring - April) a significantly reduced plasma concentration of vitamin D was found in V-LO, while in V-S (white meat consumers) this concentration was balanced with non-vegetarians. A higher incidence of deficit values was found in V-LO versus NV and V-S (67% versus 46% and 50%). It should be noted that vitamin D concentrations are low in all three groups examined (the lower limit of recommended values is 30 ng/ml), suggesting the need for supplementation or pharmaceuticals in the winter and early spring months.

3.3 Iron

Decreased utilization of minerals and trace elements from food has been observed in people with a dominant consumption of plant foods due to the high content of phytic acid in plant food (whole grains, legumes) as well as fiber (whole grains, legumes, seeds, nuts) [34]. Phytic acid and fiber form undesirable insoluble complexes with some minerals and trace elements, which cannot be used by the organism. These food commodities are significantly more consumed by vegetarians versus non-vegetarians. We observed significantly reduced serum iron concentrations in the V-LO group, hyposiderinemia occurred in 44% versus 20% in NV and 30% in V-S).

In addition to known iron deficiency disorders and diseases, the latest and lesser-known finding is that iron deficiency adversely affects the biosynthesis of long-chain n-3 fatty acids.

3.4 Iodine

The iodine content of food of plant origin is lower compared to food of animal origin due to the low iodine content of the soil. On the other hand, regular consumption of animal products (eggs, cheese, milk, meat, fish and poultry) can make a significant contribution to overall income. The literature review indicates that in vegans and vegetarians not consuming iodine supplements and seafood, iodine consumption is inadequate [35]. A wide range of mental, psychomotor and growth abnormalities cause iodine deficiency [36]. Determination of iodine is a more exact criterion than iodine intake, and urinary iodine excretion expresses the degree of saturation in the body. In group V-V we measured urinary iodine excretion 78 μg /l, in V-LO 172 μg /l and in group NV 216 μg /l. A clinically significant deficit (less than 50 μg /l) was reported in 27% V-V, 10% V-LO, but no non-vegetarian. The authors Leung et al. (14) measured median urinary iodine concentrations of 78.5 μg /l in vegans and 147 μg/l in vegetarians.

3.5 AGEs - advanced glycation endproducts

These products (AGEs - advanced glycation endproducts) are formed by the non-enzymatic reaction of an aldehyde or keto group of reducing sugars with a free amino group of amino acids, proteins, nucleic acids, phospholipids and other macromolecules; the reaction is called the Maillard reaction or glycation [37]. AGEs negatively affect the functional properties of proteins, lipids and DNA [38]. These chemical modifications accumulate in the body with age and may contribute to the pathophysiological processes associated with aging and to the complications of diabetes, atherosclerosis and chronic renal failure [39, 40]. AGEs are produced from monosaccharides (glucose, fructose), but also from dicarbonyl intermediates of the Maillard reaction, sugar autoxidation, and other metabolic pathways [41, 42]. The reactivity of monosaccharides in AGE formation is given by the ratio between the occurrence of the acyclic and cyclic forms. Only open chain sugar can enter the glycation [43]. Fructose is more reactive because it has a higher proportion of the acyclic form [44]. The mentioned processes of the Maillard reaction in the organism and exogenous AGE from food (especially culinary and technologically modified) are the main sources of intracellular and plasma AGEs [45]. CML (N^{ε}-carboxymethyllysine) values and fluorescence AGE values are significantly higher in vegetarians. CML, a major product of oxidative modification of glycated proteins, represents a common marker of oxidative stress and long-term protein damage in the aging process, atherosclerosis, and diabetes [46]. Vegetarians consume less protein and carbohydrates, and lysine intake is significantly reduced. They prefer the use of lower temperatures and shorter periods in food preparation. Some dairy products, specially cooked and with added sugar and stabilizers, have a higher CML content [47]. V-LO consumes significantly smaller amounts of dairy products (220–17 g / day versus 469–41 g/day NV, traditional nutrition) [48]. By excluding all of the above options for increasing plasma AGE values in vegetarians, we focused on monosaccharides. Fluorescence as an index of advanced glycation has been shown to increase linearly for human serum albumin incubated with glucose and exponentially using fructose [44]. The fructose-induced AGE fluorescence was higher than the glucose-induced AGE fluorescence due to the higher content of the more reactive acyclic form of fructose versus glucose.

3.6 Amino acids

The body needs to take in all the essential amino acids and in the optimal amount. Only under these conditions can amino acids from food be adequately used for protein synthesis [48]. The limiting amino acid in the protein mixture consumed (which is the lowest) is crucial for the productive utilization of all essential amino acids for anabolic processes by initiating peptide chain synthesis. Additional amino acids are incorporated into the peptide chain depending on the availability of an amount of limiting amino acid. The major limiting amino acids in plant proteins are methionine, lysine and tryptophan. The content of tryptophan is approximately the same in plant and animal proteins. Significantly reduced methionine and lysine intake in subjects with dominant or exclusive consumption of plant proteins may indicate a reduced rate of proteosynthesis, which is expressed by deficient plasma protein concentrations. Hypoproteinemia has been reported in 20% of adult vegans. The incidence of hypoproteinemia in children was higher - 33% in vegans and 11% in vegetarians. In our earlier experimental study using labeled amino acids, we found that protein synthesis was significantly reduced in young animals during the period of growth fed with plant protein (wheat gluten) compared to animal protein (milk casein). Caso et al. [49] also used the isotope technique and two diets with the same energy and protein content but with different protein sources. They reported that albumin synthesis in healthy volunteers was significantly reduced after eating plant food (67% plant protein, 33% animal protein) compared to eating 74% animal protein and 26% plant protein).

3.7 Microflora of the gastrointestinal tract

The microflora of the large intestine is an important indicator of human health. Currently, many studies point to the association of various diseases with the quantitative or qualitative representation of individual microorganisms in the gastrointestinal tract. This complex community of microorganisms can "communicate" with cells of the gastrointestinal immune system and thus coordinate the immune response to harmful pathogens. The microflora is also involved in the production of various substances such as certain vitamins or enzymes that help the proper digestion of food, but also acids, bacteriocins and the like that inhibit the growth and adhesion of harmful microorganisms to the surface of the intestinal epithelium. To ensure digestion and immune function of the intestine, it is necessary to properly represent the intestinal microflora, which is influenced by many factors, but to a large extent by the composition of the diet.

The gastrointestinal microflora consists of a very complex community of microorganisms consisting of more than 400 different bacterial species. The number of bacteria increases from 103/ ml in the stomach, 104–106/ ml in the small intestine to more than 1012/ml in the large intestine [50]. The majority of intestinal bacteria are coliform bacteria, streptococci, lactobacilli and strictly anaerobic bacteria of the genera Bacteroides, Bifidobacterium, Fusobacterium and Clostridium. Quantitatively, the microflora of the intestinal tract represents huge numbers of microorganisms approaching a trillion bacterial cells [51]. This huge proportion of bacteria makes up about half the weight of stool [52].

Up to 25–70% of diseases can be prevented by eating optimal food, its specific and balanced components. I. Mečnikov was already thinking about the connection: diet - microbes - health - disease. The bacterial flora can affect colorectal carcinogenesis by producing enzymes that transform procarcinogens into active carcinogens. These include β-glucuronidase, glycosidase, nitroreductase. The main effector mechanisms by which the microflora affects the development of cancer include: activation of procarcinogens, fermentation leading to the formation of short-chain fatty acids, formation of diacylglycerol, synthesis of pentanes and adsorption of hydrophobic molecules [53].

In our study, we divided the quantitative and qualitative representation of the microflora according to age categories and eating habits, while the numbers of microorganisms were expressed as logarithmic values. In the age category 21–30 years we found the largest differences in the total number of clostridia, in non-vegetarians this number was almost 2 orders of magnitude higher than in vegetarians, as evidenced by several studies [54]. In the case of lecithinase-positive clostridia, which play an important role in certain diseases such as e.g. We have not seen a significant difference in colon cancer in terms of eating habits. The number of spores was slightly higher in vegetarians and we assume that they enter the digestive tract mainly through plant food.

In non-vegetarians, we observed a slightly higher number of probiotic bacteria (genus Lactobacillus, Bifidobacterium), which may be due to higher intake of sour milk products in this category of people compared to vegetarians who consume these products to a lesser extent, or not at all. At the same time, we observed an increased number of staphylococci in these people (by about 2 rows), but the *Staphylococcus aureus* strain predominated in people eating plant foods. Occurrence of this strain in the gastrointestinal tract is rare, and it can enter the GIT from contaminated food or from the upper respiratory tract and cause various disease states [55]. In vegetarians, we also observed a slightly higher number (about half an order) of bacteria of the genus Enterococcus spp. and Listeria spp.

The values in the quantitative representation of individual microorganisms of the population of people aged 31–40 were mostly balanced. We observed slight differences especially in the case of eukaryotes (yeast, fibrous fungi) and clostridia, which were increased in non-vegetarians and, conversely, the numbers of listeria, spores and veilonel were higher in vegetarians. By comparing the representation of the intestinal microflora of vegetarians and non-vegetarians in the age category 41–50, we found a difference in the number of total anaerobes, which was 0.3 logarithmically higher in non-vegetarians. In contrast, the total number of aerobic microorganisms was higher in vegetarians. The number of representatives of Bacteroides spp., Veillonella spp., LP-clostridia was comparable in both groups. In vegetarians, we observed a slight increase in Enterococcus spp., Which was approximately 0.5 logarithmically higher compared to nonvegetarians. There was a small difference in the number of bacteria of the family Enterobacteriaceae, with higher values belonged to vegetarians. At the same time, we observed an increased number of Clostridium spp. about 0.5 logarithmic order. We registered the largest difference in the coagulase-positive species *Staphylococcus aureus*. The value of *S. aureus* in vegetarians reached 1.61 ± 0.38 CTU / g stool, which is significantly higher compared to the number of 0.34 ± 0.24 CTU/g stool in non-vegetarians. By comparing the group of people aged 51–60 years, we did not notice a significant difference in the amount of total anaerobes and aerobes. However, we found a significant difference in the number of LP-clostridia by a logarithmic order of magnitude higher in nonvegetarians. The production of α-toxin by LP-clostridia causes cell endothelial damage, resulting in changes that in some cases can lead to colon cancer [56]. In the age group, we registered an increased number of coagulase-positive *S. aureus*, especially in non-vegetarians.

In the etiology of colon cancer, not only genetic predisposition plays an important role, but also some microorganisms, whose quantitative and qualitative representation in the colon is largely influenced by the diet consumed. For example, some bacteria of the genus Bacteroides are able to transform certain substances into mutagens [57]. Another factor influencing this disease is the intake of carcinogenic substances that may arise during food processing [58]. Therefore, another goal was to monitor for the presence of potentially mutagenic substances in the colon. In the group of people aged 21–30 years, we recorded the presence of potentially mutagenic agents in approximately 20% of probands. In the group of non-vegetarians, we captured a slightly higher number of people (25%) compared to vegetarians (21.87%), but we did not register a significant difference between the two groups. In the age category of 31–40 years, we found a much more significant difference in the percentage of probands with detected potential mutagenic substances in the stool. In this case, it is surprising that this percentage was almost five times lower for non-vegetarians (3.84%) compared to vegetarians (19.44%). We assumed that the situation would be the opposite, as vegetarians have a higher dietary fiber intake. This is made up of important oligosaccharides that serve as prebiotics for probiotic bacteria to help prevent colon carcinogenesis. Fiber also increases the peristalsis of the large intestine, is indigestible by the human body, which mechanically "cleanses" the intestinal epithelium and removes carcinogens from the body [59]. On the other hand, people consuming a mixed traditional diet receive more frequently roasted, grilled or otherwise cooked foods in which the presence of mutagenic substances has been demonstrated [60]. However, this trend was confirmed in the category of probands aged 41–50 years, where we determined the presence of potentially mutagenic substances, especially in non-vegetarians. The percentage of the population in the age category 51–60 was relatively balanced in both groups, but at the same time almost half lower compared to the young (21–30 years) generation of probands.

Diet is one of the important indicators affecting the composition of the intestinal microflora. It is important to monitor and at the same time modulate the composition of food intake in the prevention or treatment of certain diseases. From our results of intestinal microflora values, it is clear that a randomly selected population consumes a varied diet, i. the intake of animal food is supplemented by a plant-based diet. The result of such a diet is a balanced intestinal microflora. Under certain circumstances and on the basis of certain measured parameters, it can be assessed that the conventional diet is "more advantageous" than a predominantly plant-based diet. Increased intake of beneficial bacteria in the form of probiotic products could have a positive effect on health in old age.

4. Health benefits of dominant consumption of vegetable food

4.1 Lipid parameters

Epidemiological studies document that the consumption of animal fats that contain cholesterol and saturated fatty acids causes hypercholesterolemia, while unsaturated fatty acids (plant sources) have a cholesterol-lowering effect [61]. Also, consuming a high-fiber diet prevents the risk of cardiovascular disease [7]. The hypocholesterolemic effect of fiber is explained by the binding to bile acids and the increase in fecal sterol excretion. Fermentation of soluble fiber produces short chain fatty acids that inhibit cholesterol synthesis in the liver. Whole grains, legumes, fruits, vegetables and various types of nuts are very good sources of fiber [62]. The analysis of standardized and validated dietary questionnaires on the frequency of food consumption in our vegetarians versus non-vegetarians significantly reduced the daily intake of total fats, saturated fatty acids and cholesterol and on the other hand significantly increased daily intake of vegetable fats, unsaturated fatty acids, linoleic acid and α-linolenic acid. We have found that vegetarians consume significantly more fiber, whole grains, legumes, fruits, vegetables, nuts, vitamin E, vitamin C, β-carotene and selenium. The consequence of this diet is favorable values of markers of cardiovascular risk. Total cholesterol, LDL-cholesterol and triacylglycerol levels are significantly reduced in vegetarians versus non-vegetarians and the incidence of risk-related values is low (8.2% versus 43.3% total cholesterol; 2.5% versus 28.7% LDL-cholesterol; 10.8% versus 25.6% triacylglycerols). HDL-cholesterol levels are significantly elevated in vegetarians.

4.2 Non-lipid parameters

C-reactive protein (high sensitivity, hsCRP), a marker of inflammation, can predict the risk of myocardial infarction, stroke, peripheral arterial disease, and sudden cardiac death in healthy subjects, as well as death and recurrent events in patients with acute or persistent coronary heart disease [63]. Prospective clinical studies demonstrate that atherosclerosis is not a simple disease caused only by lipid imbalance, but is an inflammatory process with highly specific cellular and molecular responses. HsCRP expresses additional prognostic value to cholesterol levels, Framingham risk score, metabolic syndrome, blood pressure with or without subclinical atherosclerosis. Vegetarians have significantly reduced hsCRP levels. No vegetarian was found to be associated with a higher risk versus 11.6%

of non-vegetarians. Favorable values of the inflammatory marker in vegetarians are explained and proven by the consumption of fruits and vegetables, which are rich sources of salicylates and other anti-inflammatory components [64]. We also measured favorable values of another non-lipid marker of cardiovascular risk in vegetarians, namely values of insulin resistance. Complex carbohydrates with a low postprandial glycemic effect, called the glycemic index, and a high fiber content are absorbed slowly and thus have a beneficial effect on glucose, insulin, insulin resistance, and blood lipids [65]. The vegetarian group versus non-vegetarian group had an IR/HOMA/value significantly reduced and very low (0.97) with no incidence of risk values versus 7.9% in the non-vegetarian group. Compared to non-vegetarians, vegetarians consume significantly more wholegrain products, legumes, barley and oat products, as well as fruits and vegetables, which contain complex carbohydrates and fiber.

4.3 Vegetable proteins

Consumption of plant proteins can also reduce the risk of cardiovascular disease. Vegetable proteins versus hen egg reference protein or other animal proteins have a reduced content of some essential amino acids and an increased content of some non-essential amino acids [66]. Consumption of essential amino acids methionine and lysine is significantly reduced in vegetarians compared to the traditional diet, intake of non-essential amino acids arginine, glycine and serine is significantly higher, when evaluating questionnaires in both types of subjects with the same protein intake corresponding to OVD, consumption is significantly increased in vegetarians alanine. Essential amino acids are relatively more effective at releasing insulin, while non-essential amino acids arginine and pyruvate precursors are effective at secreting glucagon. Glucagon increases (and inhibits) c-AMP-dependent mechanisms that suppress fat and cholesterol synthesis enzymes and, conversely, increase LDL receptor activity in the liver [67]. The effect of a chronic increase in glucagon activity by regular, sufficient consumption of plant proteins means a reduction in de novo lipogenesis, a decrease in fat stores, a reduction in cholesterol and LDL-cholesterol synthesis, and a reduction in triacylglycerol synthesis [68]. These literature data suggest that the favorable lipid profile and low IR (HOMA) value in vegetarians may also be due to higher intake of the non-essential amino acids arginine, glycine, alanine and serine from plant protein consumption.

4.4 Antioxidant vitamins; oxidative damage products of lipids, proteins, DNA

Oxidative stress can lead to cell dysfunction and eventually cell death. It is defined as an imbalance between pro-oxidants or free radicals on the one hand and antioxidant systems on the other. The harmful effects of oxygen occur through the production of free radicals that are toxic to cells (superoxide anion, hydroxyl radical, peroxyl radical, hydrogen peroxide, hydroperoxides and peroxynitrite anion) [69]. Subjects with a dominant consumption of plant foods have significantly increased plasma concentrations of antioxidant vitamins C, E, β-carotene, values for lipids standardized required E as well as the value of the ratio of large C and E and these average values of antioxidants are above the threshold. The incidence of protective (above-threshold) values is high in vegetarians - 92% versus 42% for vitamin C, 67% versus 33% for vitamin E, 100% versus 79% for vitamin C/vitamin E, 87% versus 50% for vitamin E /cholesterol, 96% versus 62% for vitamin E/triacylglycerols and 67% versus 17% for β-carotene. The results document the better antioxidant status of vegetarians as a consequence of regular and sufficient consumption of protective food and are consistent with the results of other authors [70]. Due to more effective antioxidant protection, we measured significantly reduced plasma values of lipid peroxidation (conjugated fatty acid dienes as the first product of the process), insignificantly reduced values of protein damage and significantly reduced values of oxidized purines and oxidized pyrimidines in lymphocytes in vegetarians. The values of oxidative damage of DNA, lipids and proteins were significantly lower in persons with above-threshold values of antioxidant vitamins compared to sub-threshold, deficient values.

5. Conclusion

Epidemiological data clearly suggest that high and most important regular consumption of fruits, vegetables, dark or whole grains, cereal germ, vegetable oils and oilseeds rich in minerals and trace substances, mono- and polyunsaturated fatty acids, antioxidant vitamins, fiber, complex carbohydrates, flavonoids and nutrients together with a healthy lifestyle protect against degenerative diseases. Our results indicate that vegetarian nutrition can represent effective disease prevention. Optimal traditional nutrition with sufficient consumption of protective food commodities can also ensure beneficial effects on health.

Acknowlegment

This study was supported by research agency VEGA under the contract 1/0096/17 and APVV-16-171.

References

[1] Gusak, M.A., DellaPenna, D.: Improving the nutrient composition of plants to enhance human nutrition and health. Annu.Rev.Plant Physiol. Plant Mol.Biol.,50, 1999,133-161.

[2] Allen, L.H.: Causes of vitamin B12 and folate deficiency. Food Nutr. Bull.,29, 2008,20-34

[3] Leitzmann, C.: Vegetarian diets – what are the advantages? Forum Nutr.,57, 2005,147-156.

[4] Craig, W.J.: Health effects of vegan diets. Am.J.Clin. Nutr.,89,2009,1627-1633.

[5] Elmadfa, I., Singer, I.: Vitamin B12 and homocysteine status among vegetarians, a global perspective. Am.J.Clin.Nutr.,89,2009,1693-1698.

[6] Sabaté, J.: The contribution of vegetarian diets to human health. Forum Nutr.,56, 2003,218-220.

[7] Rajaram, S., Sabaté ,J.: Health benefits of a vegetarian diet. Nutr.,16,2000,531-533.

[8] Craig, W.J., Mangels, A.R.: Position of the American Dietetic Association, vegetarian diets. J.Am.Diet. Assoc.,109,2009,1266-1282.

[9] Erkkila, A., DeMello, V.D., Risérus, U. et al.: Dietary fatty acids and cardiovascular disease, an epidemiological approach. Prog.Lipid Res.,47,2008,172-187.

[10] Key, T.J., Fraser, G.E., Thorogood, M.: Mortality in vegetarians and nonvegetarians, detailed findings from a collaborative analysis of 5 prospective studies. Am.J.Clin. Nutr.,70,1999,516-520.

[11] Mann, J.I., Appleby, P.N., Key, T.J. et al: Dietary determinants of ischaemic heart disease in health conscious individuals. Heart,78,1997,450-455.

[12] Fraser, G.E.: Associations between diet and cancer, ischemic heart disease and all-cause mortality in non-Hispanic white California Seventh-day Adventists. Am.J.Clin.Nutr.,70,1999,532.

[13] Hu, F.B.: Diet and lifestyle influences on risk of coronary heart disease. Curr. Atheroscl. Rep.,11,2009,257-263.

[14] Szeto, Y.T., Kwok, T.C., Benzie, I.F.: Effect of a long-term vegetarian diet on biomarkers of antioxidant status and cardiovascular disease risk. Nutr.,20,2004, 863-866.

[15] Krajcovicova-Kudlackova, M., Babinska,K., Valachovicova, M.: Health benefits and risks of plant proteins. Bratisl.Med.J.,106,2005,231-234.

[16] Caroll, K.K., Kurowska, E.M.: Soy consumption and cholesterol reduction, review of animal and human studies. J.Nutr.,125,1995,594-597.

[17] Čerhata, D., Baureová, A., Ginter, E.: Stanovenie askorbovej kyseliny v krvnom sére vysokoúčinnou kvapalinovou chromatografiou a jeho korelácia so spektrofotometrickým stanovením. Čes. Slov. Farm., 43, 1994, 166-167.

[18] Lee, B. L., Chua, SC., Ong, H. Z., Ong, H. Y.: High performance liquid chromatographic methods for routine determination of vitamins A and E, and β- carotene in plasma. J. Chromatogr., 581, 1992, 41-43.

[19] Talwar, D., Ha, T. K. K., Cooney, J., Brownlee, C., O'Reilly, D. S.: A routine method for the simultaneous measurement of retinol, α-tocopherol and five carotenoids in human plasma by reverse phase HPLC. Clin. Chim. Acta, 270, 1998, 85-100.

[20] Wong, S. H. Y., Knight, J. A., Hopfer, S. M., Zaharia, O., Leach, C. H. N., Sunderman, F. W.: Lipoperoxidates in plasma as measured by liguid chromatographic separation of malondialdehyde- thiobarbituric acid adduct. Clin. Chem., 33, 1987, 214-220.

[21] Levine, R. L., Garland, D., Oliver, C. N., Amici, A., Climent, I., Lenz, A. G., Ahn, B. W., Shaltiel, S., Stadtman, E. R.: Determination of carbonyl content in oxidatively modified proteins. Methods in Enzymol., 186, 1990, 464-478.

[22] Collins, A. R., Dušinská, M., Gedik, C. M., Stetina, R.: Oxidative Damage to DNA: Do have a reliable biomarker? Environ. Health Perspect., 104, 1996, 465-469.

[23] Collins, A., Dušinská, M., Franklin, M., Somorovská, M., Petrovská, H., Duthie, S., Panayiotides, M., Rašlová, K., Vaughan, N.: Comet assay in human biomonitoring studies, reliability validation and application. Environ. Mol. Mutag., 30, 1997, 139-146.

[24] Dušinská, M., Collins, A.: Detection of oxidised purines and UV-induced photoproducts in DNA of single cells, by inclusion of lesion-specific enzymes in the comet assay. ATLA , 24, 1999, 405-411.

[25] Munch, G., Keis, R., Wessels, A., et. al. Determination of advanced glycation end products in serum by fluorescence spectroscopy and competitive ELISA. Eur. J. Clin. Chem. Clin. Biochem., 1997, 35, 669-677.

[26] Sandell, E. B., Kolthoff, I. M.: Micro determination of iodine by a catalytic method. Microchemica Acta: 1, 1937, 9-25.

[27] Mitsuoka, T., Hayakawa, K. Die Zusammensetzung der Faecalflora der verschiedenen Altersgruppen. Die Faecalflora bei Menschen. I, 1972, 333-342

[28] Krajčovičová-Kudláčková, M., Blažíček, P., Kopčová, J. et al.: Homocysteine levels in vegetarians versus omnivores. Ann.Nutr. Metab.,44,2000,136-138.

[29] Varela-Moreiras, G., Murphy, M.M., Scott J.M.: Cobalamin, folic acid, and homocysteine. Nut.Rev.,67,2009,69-72.

[30] Krajcovicova-Kudlackova,M., Blazicek, P., Mislanova C. et al.: Nutritional determinants of plasma homocysteine. Bratislava Med.J.,108,2007,510-515.

[31] Rasmussen,K., Moller,J.: Total homocysteine measurement in clinical practice. Ann.Clin. Biochem.,37,2000,627-648.

[32] Chan, J., Jaceldo-Siegl, K., Fraser, G.E.: Serum 25-hydroxyvitamin D status of vegetarians, partial vegetarians, and non-vegetarians: the Adventist Health Study-2. Am.J.Clin. Nutr.,89,2009,1686-1692.

[33] Crowe, F.L., Steur, M., Allen, N.E. et al.: Plasma concentrations of 25-hydroxyvitamin D in meat eaters, fish eaters, vegetarians and vegans: results from the EPIC-Oxford study. Publ.Health Nutr.,14,2011,340-346.

[34] Hunt, J.R.: Bioavailability of iron, zinc, and other trace elements from vegetarian diets. Am.J.Clin. Nutr.,78,2003,633-639.

[35] Lightowler, H.J., Davies, G.J.: Sources of iodine in the vegan diet. Proc. Nutr.Soc.,55,1996,13.

[36] Delange, F.: The disorders induced by iodine deficiency. Thyroid,4,1994, 107-128.

[37] Njoroge, F.G., Monier, V.M.: The chemistry of the Maillard reaction under physiological conditions, a review. Prog.Clin.Biol. Res.,304,1989,85-107.

[38] Ledl, F., Schleicher, E.: The Maillard reaction in food and in human body, new results in chemistry, biochemistry and medicine. Angew.Chem.Int. Ed.Engl., 29,1999,565-706.

[39] Makita, Z., Bucala, R., Rayfield, E.J. et al.: Reactive glycosylation end products in diabetic uremia and treatment of renal failure. Lancet,343,1994,1519-1522.

[40] Vlassara, H., Bucala, R., Stricker, L.: Pathogenic effects of AGEs, biochemical, biologic and clinical implications for diabetes and aging. Lab. Invest.,70,1994, 138-151.

[41] Thornalley, P.J.: The glyoxylase system, new developments towards functional characterization of a metabolic pathways fundamental to biological life. Biochem.J., 269,1990,1-11.

[42] Wells-Knecht, K.J., Zyzak, D.V., Litchfield, S.R. et al.: Mechanism of autoxidative glycosylation, identification of glyoxal and arabinose as intermediates in the autoxidative modifications of proteins by glucose. Biochem.,34,1995,3702-3709.

[43] Beswick, H.T., Harding. J.J.: Aldehyds or dicarbonyls in non-enzymatic glycosylation of proteins. Biochem.J.,226,1985,385-389.

[44] Jakuš,V., Rietbrock, N., Hrnčiarová, M.: Study of inhibition of protein glycation by fluorescence spectroscopy. Chem Papers,52,1998,446.

[45] Takeuchi, M., Makita, Z., Bucala, R. et al.: Immunological evidence that non-carboxymethyllysine advanced glycation end products are produced from short chain sugars and dicarbonyl compounds in vivo. Mol. Med.,6,2000,114-125.

[46] Drusch, S., Faist, V., Erbersdobler, H.F.: Determination of N^{ε}-carboxymethyllysine in milk products by a modified reversed-phase HPLC method. Food Chem.,65,1999, 547-553.

[47] Krajčovičová-Kudláčková,M., Šebeková,K., Schinzel,R., Klvanová,J.: Advanced glycation end products and nutrition. Physiol.Res.,51,2002,313-316.

[48] Krajčovičová-Kudláčková, M.: Health risks of alternative nutrition. Klin.Biochem. Metab.,9,2001,187-193.

[49] Caso, G., Scalfi, L., Marra, M. et al. : Albumin synthesis is diminished in men consuming a predominantly vegetarian diet. J.Nutr.,130,2000,528-533.

[50] Maratka, Z.: Klinická gastroenterologie. Praha: Avicenum, 1988. 322-329.

[51] Savage, D.C.: Microbial ecology of the gastrointestinal tract. *Ann Rev Microbiol* 1977, 31, 107-133.

[52] Stephen, A.M., Cummings, J.G.: The microbial contribution of human faecal mass. *J Med Microbiol* 1980, 13, 45-56.

[53] Kuchta, M., Pružinec, P. a kol.: Probiotiká, ich miesto a využitie v medicíne. Bratislava: Bonus CSC, ISBN 80-968491-7-4.

[54] Mueller, S., Hanisch, S.Ch., Norin, E., Alm, L., Midtved, T. a kol.: Differences in Fecal Microbiota in Different European Study Populations in Relation to Age, Gender, and Country: Cross-Sectional Study. *Appl Environ Microbiol* 2006.

[55] Stiles, M.H., Chapman, G.H.: Probable pathogenic *Streptococci* and *Staphylococci* in Chronic Low Grade Illness, An Analysis of Their Frequenci in Three Hundert and Ninety-five Cases. *Arch Otolaryngol* 1940, 458-466.

[56] Lye, H. S., Rusul, G., Liong, M. T.: Removal of cholesterol by lactobacilli

via incorporation and conversion to coprostanol. *J Dairy Sci* 2010, 93, 1383-1392.

[57] Knasmuller, S., Kassie, F., Lhoste, E.F., Bruneau, A., Ferk, F., Zsikovits, M.: Effect of intestinal microfloras from vegetarians and meat easters on the genotoxicity of 2-amino-3-methylimidazo[4,5-f]quinoline, a carcinogenic heterocyclic amine. *J Chromatography* 2004, 211-215.

[58] Beije, B., Moller, L.: 2-nitrofluorene and related compounds - prevalence and biological effect. *Mutatu* 1988, 196, 177-209.

[59] Kment, M., Jirásek, V., Frič, P.: Gastroenterologie. Praha: Karolinum, 1999. 277-304.

[60] Sugimura, T., Nagao, M., Kawachi, T., Honda, M., Yahagi, T., Seino, Y. a kol.: Mutagen - carcinogens in food, with special reference to highly mutagenic pyrolityc products in broiled foods. Origins of Human Cancer, Cold Spring Harbor Laboratories. Cold Spring Harbor NY 1977, 1561-1579.

[61] Mattson, F.H., Grundy, S.M.: Comparison of effects of dietary saturated, monounsaturated, and polyunsaturated fatty acids on plasma lipids and lipoproteins in man. J.Lipid Res.,26,1985,194-202.

[62] Kris-Etherton, P.M., Yu-Poth, S., Sabaté, J. et al: Nuts and their bioactive constituents, effects on serum lipids and other factors that affect disease risk. Am.J.Clin.Nutr.,70,1999,504-511.

[63] Bassuk, S.S., Rifai, N., Ridker, P.M.: High sensitivity C-reactive protein, clinical importance. Curr.Probl. Cardiol.,29,2004,439-493.

[64] Lawrence, J.R., Peter, R., Baxter, G.J. et al: Urinary excretion of salicyluric and salicylic acids by non-vegetarians, vegetarians and patients taking low dose aspirin. J. Clin.Pathol.,56,2003,649-650.

[65] Reaven, G.M.: Diet and syndrome X. Curr.Atheroscl.Rep.,2,2000,503-507.

[66] Krajčovičová-Kudláčková, M., Babinská, K., Valachovičová, M.: Health benefits and risks of plant proteins. Bratisl.Med.J.,106,2005,231-234.

[67] McCarty, M.F.: Vegan proteins may reduce risk of cancer, obesity, and cardiovascular disease by promoting increased glucagon activity. Med. Hypotheses, 53,1999,459-485.

[68] Rudling, M., Angelin, B.: Stimulation of rat hepatic low density lipoprotein receptors by glucagon. J.Clin.Invest.,91,1993,2796-2805.

[69] Bonnefoy, M., Drai, J., Kostka, T.: Antioxidants to slow aging, facts and perspectives. Presse Med.,31,2002,1174-1184.

[70] Szeto,Y.T., Kwok,T.C., Benzie,I.F.: Effects of a long-term vegetarian diet on biomarkers of antioxidant status and cardiovascular disease risk. Nutr.,20,2004,863-866.

Chapter 5

Vegetarianism and Veganism: Conflicts in Everyday Life

Abstract

In everyday situations, the experience of being a vegetarian or a vegan occurs within a process of conflict and practices of negotiation involving decisions, refusals, consumption acts, and proximity and distance between people in their relationships, mainly including the family. Many dilemmas result from the inconsistency between theory and difficult practices to be obeyed. To understand how this phenomenon, the chapter uses the interviews with vegetarians considering different alimentary restrictions and data obtained from observation in virtual groups of vegan activists. We have conducted the research between 2015 and 2017 as part of a larger project entitled: The Social Place of Animals in Contemporaneity.

Keywords: vegetarianism, veganism, daily life, eating habits, animal rights

1. Introduction

The acts of eating and choosing the diet transcend the demand of nutrients. Though eating is a need for the living body maintenance, it is, fundamentally, a social fact that prescribes what must be eaten, and when, how much, and how.

Eating habits are cultural goods that may identify a nation, a region, a group. What ones eat translate a feeling of cultural belonging as well as communion.

In complex and fragmented societies, the identity may be related to a lifestyle, not necessarily linked to the relations of production, but related to the belonging in groups that share some elective affinity that, through consumption, communicate the corresponding worldview to others.

This chapter aims to consider about how the construction of identity based on the denial of meat consumption and on the adoption of other types of food that end up translating a lifestyle, shortly understood as a distinctive one, shared by others, and a guide to a meaningful behavior. I am specifically talking about vegetarian and vegan people and the contradictions they find in their everyday lives.

As Douglas and Isherwood [1] sustain, goods give marking services, intrusion, exclusion, and the consumption classifies and organizes the world, as part of the cultural system. According to the authors, including consumer goods, even the trivial ones, serve to this meaning, like dance and poetry. Likewise, Featherstone affirms that the consumption is fundamentally a producer of signs, despite consumer goods values of use [2].

From this perspective, I intend to comprehend the adoption of a vegetarian diet and the everyday tensions that vegetarian people, vegans and critics of this diet, despite the internal conflicts that they live at least during their basic meals on a daily basis. The thesis sustained is that this tension does not refer to only consumption divergences, but to what consumption represents: it is about the organization and sense of the world in conflict.

The motivations of the individuals on their decisions of consumption can also be superposed. In general, studies on consumption classify three tendencies that superpose historically and we can notice in empirical studies: (1) consumption by distinction – they used to believe that consumption worked as social distinction among social classes. Lower classes used to imitate, possibly because of envy, the consumption among higher classes that used to modify their consumption so the distance remained visible. (2) hedonistic consumption – a kind of consumption in which the distinction matters less or barely nothing, since the value is the individual pleasure with no need of ostentation. (3) consumption ethically motivated – ethics above pleasure and health and, therefore, above the individual.

In this chapter, we are analyzing the motivations, conflicts and contradictions among those who adopt a vegetarian diet or a strict vegetarian one, also known as vegan. To achieve this goal, we have interviewed ten people that are self-declared vegetarian and vegan based on an open script. We have conducted most of the interviews in Cuiabá, state capital of Mato Grosso, Brazil, as well as other cities in Brazil. We have incorporated these spontaneous testimonies and informal talks to this research. We have kept the anonymity of all participants. In order to complement the analysis, we have also followed virtual group discussions. The results vouch for the existence of sociability conflicts in general, with particular reference to the family, but also reveal internal conflicts in which the individuals question their own practice, the reach of their option when it comes to animal protection or the environment and, above all, the difficulty in obtaining coherence between the theory and the practice.

2. About vegetarians

The abstinence of meat consumption and animal source foods, may it be total- or partially, is the element of some religious practices as Buddhism and Seventh-day Adventist Church [3]; there is rejection of pork by Jews and Muslims and of beef in India [4]. Others opt for a secular vegetarianism, free of religious motivations [5].

The contemporary society provided the creation, diffusion, and resignification of restrictive diets that appear regardless of a religious belonging stricto sensu even considering a cultural heritage that leans to the habit of eating or rejecting meat or vegetables. However, it is likely to be connected to a more wide understanding of spirituality and reconnection to nature, as seen in some new era speeches, or might as well be or not be linked to groups whose coefficient of belonging is, sometimes, subtle, as in virtual communities.

According to Whorton [5], vegetarianism has grown because of moral and social tendencies based on appropriated precepts on the mystic from the Orient, and has created a relation between the neglecting of meat consumption and the demand for peace, with a concern related to environmental crisis and the demand of body health.

As stated by Beardsworth and Keil [4], vegetarianism is sustained by the interrelation of beliefs, attitudes, and nutritional practices, and the vegetarians[1] are converted after close and critical examination of their diets until then. Therefore, their practices are the results of processes of reflection and opposition to what they have culturally received. In a similar way to the ones presented by Beardsworth and Keil, with the adoption of a specific diet, it becomes possible to see that the vegetarian diet is more related

[1] In this case, it is about people who converted themselves to vegetarianism and not those who were socialized to vegetarianism since their childhood.

to individual experiences and to wishes that have been built more reflexively from information than acceptance, either authoritative or not, from shared group codes.

Claiming to be vegetarian has a meaning that, under the risk of misinterpretation, cannot be aprioristically considered. Its meaning is given and renewed on a daily basis. Generally, vegetarianism is a staple diet that abolishes meat or the one that consists of an exclusive vegetable-based diet. Vegetarian practices are more plural and do not negatively merge in a way to avoid the consumption of specific products, but in the construction of consumption habits of other products, in the discovery and invention of new recipes that can even be inspired or imitate recipes that take meat on their preparation.

Beardsworth and Keil [4] classify six types of vegetarian diets according to a set of feeding practices that vary on a scale from lower to higher strictness. Down to the less strict side, there are those who may eventually have some meat and, in general, the white ones. The second type includes those who accept fish; in the third one are those who consume eggs, milk and other dairy products, followed by those who may have some dairy as long as they do not contain any derived product from the slaughter, such as rennet. Up to the strictest side are those who do not eat any animal products. Conforming to these authors, it is important to identify the types of vegetarian diets, but they highlight that their participation is not permanent in each of the categories. Individuals move along the scale both ways until abandoning the category.

Among those who we have interviewed, all self-declared vegetarians, none claims to consume any kind of meat, although there might be times when they suspend their diet. However, some of them declared they know people who claim to be vegetarian but eat fish or white meat. Some of them might accept eating dishes prepared with meat, others reject any contact with the animal origin, including their handling. One mentioned the discomfort of using cutlery that had previously been utilized with meat even though they had also been washed up.

One of the participants, besides being a vegetarian, claims to have a macrobiotic diet and another one claims to be vegan, and refuse to consume animal products of any source and not just regarding food. The others refuse to eat meat, but they accept eggs, milk, and other dairy products, and among these, some of them manifested against the leather and the animal testing industry, even though they still consume them.

The reactions of disgust or indifference towards meat and/or animal products and the decisions of what one puts or does not put on the plate and what one takes or does not take to their mouth taken at least three times a day and every day are different. Each one of them explains the reasons why they have joined the particular diet. None of them was raised in a vegetarian diet and they have decided to join it at their adult stage. Their reasons and their corresponding lifestyle are various, though not excluding, and they mix ethical impulses concerning animals as well as demand of health.

Their motivations differ on the rank of importance and on each one's life among the participants. It shows that even though they did not have an ethical impulse at first, they end up having it through the course of their lives and it becomes related to an essential aspect when having to justify their lifestyle. Considering the testimonies we have analyzed, the protection of animals appears as the top factor, even more than the demand of health, for either participation or maintenance of their diet. This fact is at least curious once health has become a highly valuable capital in the contemporary culture.

Fox and Ward [3] have studied the motivations that led youngsters to the conversion into vegetarianism mainly in The United Kingdom, Canada and The United States. They have noticed that the decision of a diet without meat, the fight against animal abuse and the worry with personal health are the main elements cited as encouraging, but they have also listed items related to disgust when eating meat, the association with patriarchy, their friends' beliefs, and family influence.

Concerning health, this motivation seems to preponderate among partial vegetarians, the ones who do not eat red meat or only fish or those who consume organic products. The vegetarians have been classified into two main types concerning their motivations: the "health vegetarians" and "the ethical vegetarians"; however, participants of both groups also practice lacto-ovo vegetarianism and end up, during this process, joining a semi-vegan diet.

> *I decided to stop eating meat because I wanted to be healthier. I have read some books about natural food that said meat would rot in our stomachs. I felt disgust and decided to stop eating meat so I would be healthier. At that time, I cut off everything, including white flour, white rice, white sugar, and soda. Gradually, I started eating some things, back to my old habits, but I could never eat meat again. I have not even tried, have I? I had not craved meat because it was still disgusting to me. After years, I started thinking about environmental and animal issues. Today, I politically defend vegetarianism and I want to become a vegan. Doing without cheese is still quite difficult though. (Woman 1).*

For this participant, a young self-employed woman, and in other similar testimonies, when I asked which factor is the most important so they would keep in their diet, animal protection appeared more than demand of health, again for either participation or maintenance of their diet.

According to Lipovetsky [6], we live in a new phase of consumption that represents the society in which he calls hypermodern, among other things, as a time of medicalization of life and consumption. Health is a responsibility of social actors.

In the testimonies we have analyzed, the ethics regarding animal life is the biggest reason for those who keep being vegetarians. Two of them abandoned their vegetarian diets because of health issues and are, nowadays, omnivorous and critics of the vegetarianism they had adopted at a stage of their lives, which partially confirms Lipovetsky's diagnosis that the *homoconsumericus* is giving its place to the *homosanitas*. [6].

Once we clarify the heterogeneity of the diets within the vegetarian label and the motivations that reveal this multifaceted character of contemporaneity, we start discussing everyday conflicts due to the diet, even though we do not ignore the peacemaking feeling provided by food.

Changing food patterns has an effect on social relations, mainly the family ones, but also on friend's network; by converting into vegetarianism one can find sympathy and support or even criticism, confusion, and hostility, as attested by Beardsworth and Keil [4], who have realized, among their respondents, the contrast between acceptance and criticism. This last one appears to be more emphasized concerning parents' reactions given their children's conversion. Mothers seem to be more sympathetic and tolerant about the conversion.

The first conflict revealed in the respondents' speeches happened in their families. Besides the transformations observed in the contemporary family, the mom is traditionally the one responsible for her children's first socialization and it is about her role that the obligations to feed the family lie. It also carries her activities of affection and anxiety, once the mother is the one whose success of socialization depend on. Mothers feed their children according to a rule of society in terms of what, how, and how much to eat.

A disagreement related to this manifested learning when refusing appropriate food and the intake of other inappropriate food cause some family conflicts. This is precisely what we see in the relations between a vegetarian and a non-vegetarian family. This individual becomes a disruptive element of family tradition, of union during meals and communion of values. The refusal of the shared dish is seen as a refusal of ideas as well as the family's worldview.

Beardsworth and Keil [4] affirm that as vegetarianism can involve a rejection of the food that parents offer, such practice can be understood as rejection to their own parents. According to these authors, several family occasions are turned into tension occasions, and the most critical one is Christmas, given the importance of this celebration for the maintenance of the family identity. The tension exists either when vegetarians visit their families or when relatives visit vegetarian families. Furthermore, the situations of conflict are less common when one member of the couple is vegetarian; they also observe that vegetarian couples tend to hold on to each other against the rest of the family.

Two of the participants, declared as middle-class ones, mentioned Christmas specifically and spontaneously when questioned about situations of conflict. One of them, a woman, said she feels a bit excluded of this festivity, when she would never share the main course, even though she had a very strong participation in the arrangements of the party, with typical abundance and exaggeration promoted by her mother. The other one, a man, said he would starve in these occasions for meat would be in every single dish, even in the salads. During the Holy Week, it would not be different because the only dish was the "*bacalhoada*," a codfish dish popularly consumed at this time of the year.

As he cannot cook, the others would commit him to their choices, which did not consider his restriction. The decision of not sharing the so-called appropriate dish may become a non-sharing of habits, ideas, and worldviews. Likewise, the refusal of an offered dish might be read as insubordination to rules of family relationship.

As an example, we have the testimony of a woman, omnivorous, 45 years old. Her only child has become a vegetarian at the age of 20, influenced by friends. She confirmed she did not understand her motivations and feared for her health. Through the testimony, which she participated spontaneously when aware of this research, she mentioned several situations of arguments and fights. She claimed to feel rejected by her kid's rejection of her motherly food. By refusing not eating a dish prepared by her, she used to feel rejected affectionately and the non-consumption tended to become non-affection.

Mauss [7], when studying human transactions through the analysis of ethnographical exchanges in Polynesia, Melanesia, and the American Northwest, realizes that the gift-exchange demands three obligations: giving, important for building reputation; receiving; and reciprocating. In this sense, we can conclude that giving is a fundamental social action for gratitude and, hence, receiving prestige. Receiving would mean a representative action of acceptance of an alliance, while the refusal would mean an affront.

This idea contributes to understanding the feeling of a mother when her child refuses her food for feeling offended by the offer of meat, which represents indifference regarding ideological option. This is all about linguistic incompetence.

Beyond the family situation, conflicts rise in other sociability loci, mainly when the imagination of omnivorous people of what vegetarian people should be is not compatible with the actual practice.

The stereotype of vegetarian people pictures individuals linked to alternative movements, intention to become healthier, eastern religions, concerns about nature and animals above all, according to the participants in this research. Linguistic incompatibilities lie on face-to-face relations, when one presumes the other one is not a vegetarian and, by acknowledging their food choice, one presumes the ideal type of a vegetarian.

One of the participants, a former vegetarian, affirms that "vegetarian people must eat only vegetables" and her sister-in-law is "this big" (she says making hand gestures meaning overweight or obesity) and only eats "pasta and cheese." When I asked about her sister-in-law's motivation, this participant affirms she is an advocate of animal causes. In this case, it becomes possible to realize the trouble there is between motivation and expectation of how vegetarians should actually be when seen by others. When imperatively affirming, "Vegetarians must eat vegetables" it is possible to see the attempt to establish the other's consumptions based on what they expect or imagine. It is, thus, the imposition, or effort, of some sort of consumption. There is an idea of an appropriate consumption of a specific social category and they assume that that category consumes, as premises of subjective and identity construction, specific products and specific bodies that are considered validators of this very category.

In another testimony, a 21-year-old vegetarian woman tells:

> *Well, if you really want to know, people always want to tell me what to or not to eat. Besides the usual campaigns for me to eat meat, I remember one trip when I ate one of those cakes made of black-eyed beans. I asked many times if there were not any animal products and chose the fried one. We were able to choose between the fried or the baked one. I chose the fried one, without filling, because it was a shrimp filling. Next to me, a woman started laughing, saying... hahaha she is a vegetarian and eats the fried cake – I do not remember its name – hahaha. She found it absurd that I was eating something fried and being a vegetarian. The oil where she fried the cake was a vegetable one. It drives me crazy. I used to smoke cigarettes and people kept telling me: why, you say – even though they kept telling me I claimed to be vegetarian and was not vegetarian – then, you say you are a vegetarian and smoke? I would not eat the cigarettes and it was made of vegetables. People thought I had to be green and healthy. I just felt like not eating meat. (Woman 2)*

Therefore, this claimed identity clashes with a kind of mental construction of a stereotyped individual of normative behavior. There is this prescriptive idea of how someone must be and what a vegetarian must consume as mentioned above: "Vegetarians must eat vegetables" or "She says she is a vegetarian, but she only eats cheese and pasta." The obese body, smoking habits and not so healthy food habits affront the expectations of what a vegetarian should look like.

> *I was at a restaurant with someone who eats meat and we ordered a cheese skewer for me and a meat skewer for the person who was with me. As a side dish for the meat skewer, there was rice and cassava. I asked the waiter to bring the rice on the side so the meat gravy would not dirty the rice and we both were sharing this one. He said the meat would not come with gravy. I said I knew it, but I still would like it on the side, just in case. He said there was no gravy with the meat. It was hard convincing him that it happens that meat, with no gravy, still has its juice when resting, this juice would dirty the rice, and then I would not be able to eat it. He said that eating rice and cheese made no sense to him. (Woman 3)*

Ideas of pollution only make sense when in Ref. to a total structure of thought, according to Douglas [1]. As for the waiter cited in the testimony, the meat would not dirty the rice given the idea the meat was clean. On the other hand, according to the woman, the meat was a pollution agent when it touched the rice. Intuitively, the waiter realizes the symbolic character of food by demanding the sense of matching cheese and rice. This sense is only understood in a cultural system, for its foundation is not on reason, which is something even more complex in a multifaceted contemporary society.

Attempts to explain it are not enough and they are countless. Explanations do not serve as an interpretation because they are not coherent or comprehensive. Douglas demonstrates that the "abominations" of Leviticus refer to ambiguities, that is to say, the abominations lie on what challenges a socially built logic. Everything that is not in accordance with the structure of classification in the culture in question is considered ambiguous or anomalous and, as these, unclean.

I have had a very unpleasant situation. My daughter and I are vegetarians. My daughter, a child, went to a friend's house, they were making barbecue, and my daughter explained them that she does not eat meat. Her friend's father said she could help herself with some grass from the yard. She got very sad and I found the father's comment was quite offensive. I swore that when his daughter would come to my place, I would offer my dog for her to eat, since she could not do without meat. (Woman 4).

The intention is to offend what is different. It is clear that that man, by offering some grass from the yard, is animalizing the other. The mother hypothetically revenges the situation when she suggests the offer of her dog as food. A vegetarian does not eat grass and an omnivorous will not eat a pet dog. Everyone knows that, but they use these allegations and offers with the only purpose of offending others. It maintains and reinvigorates the belief in superiority of options on which the identity is built.

However, it is important to underline that the identity, in this perspective, is built almost as experimentally. Individuals start conceiving and noticing how far they might go, and what brings them satisfaction, in terms of craving, and not need, when they face multiple restrictions, choices, learning about new dishes, new restaurants, points of sale, recipes, relapses, and new restrictions.

For me, becoming a vegetarian was part of a long process. I have been a vegetarian for ten years. I have been defining myself as a vegetarian all this time, but actually, I believe I am more vegetarian now than back then, if I can say that. Well, I am lacto-ovo vegetarian and I have already been questioned about the fact that I say I am a vegetarian. I answer that it is, like, an abbreviation of my eating habit: the ovo-lacto vegetarianism. As I said, I used to eat feijoada (a typical Brazilian dish made of black beans, pork sausages and other cuts of pork). I used to take out the meat and eat the beans. When it was meat and potatoes, I used to eat the potatoes. Not today, I do not even try it if they were cooked together. I look at it as if I were looking at a shoe, a chair. It is an object and not food. In my opinion, food is something else. (Woman 2)

In contrast, there are those who have tried assorted diets, macrobiotic and vegetarian ones, and then returned to their omnivorous diet, even if this one was not the same way they were raised. That demonstrates the construction of omnivorous individuals is equally processual and reflexive, although it may seem natural in different speeches.

This multiplicity of comings and goings-away and dietary, gustatory, and social experiences mark individual biographies, which constitutes kaleidoscopical individuals. Dietary values are worth as an allegation, they just have rhetorical value, given the option one ingests or loathes, as they do not base food on nutrients, but a tangle of mental constructions, social representations, and personal idiosyncrasies.

The abominations of certain food can justify the protection of the body, though the adjacent objective is the maintenance of the social organism. Consequently, dietary rules extrapolate their practical aspect and are part of a symbolic system where there is a dispute of advantages and disadvantages of diets based, recurrently, on the three most popular cultural authorities: religiosity, nature and science.

2.1 Religiosity

The analysis conducted by Douglas [1] shows that behind an apparent rationality of Jewish dietary rules, there is a complex symbolic system and it demonstrates that human acts are influenced by a lot of things beyond rationality and medical criteria, which helps us realize that food does not just feed, but it is part of the establishment of identities. Besides this allegation of the humanity omnivorous nature, vegetarians are asked about the religious prescription that advocates that God created animals for human consumption:

This is one of the most annoying topics. People come to me and tell me I must eat meat because God said so. I reply, so what? I am not killing an innocent animal because God or the devil said so. Then they call me atheist, as if it were a flaw. The curious thing is, I remember now, is that that astonished woman who told me that was a divorcée. Does God approve the divorce? People follow God, the bible, or the law according to what is convenient and they think they have the right to criticize who decides to live by other principles. (Man 1)

Coherence is being demanded in this example. A religious individual should, according to this testimony, follow all biblical principles, including the one regarding marriage and the ingestion of meat as well as other demands. Such coherence does not exist for innumerous historical reasons that restrict the way of appropriating the testaments. On the contrary, vegetarians are charged for their coherence as well. If they defend animal rights, the opponents of the vegetarian diet discuss that one must refuse products tested on animals, including vaccines, and to the extreme living with animals considered plagues in cultivations given the impossibility of the use of pesticides and also a stimulated conviviality between predators and preys.

Even though they demand logic, neither vegetarians nor defenders of meat consumption practice it, because both deprive on the consumption of certain species or specific situations completely irrationally.

2.2 Nature

Vegetarians question the culture where they were born, but they also question the humanity omnivorous nature. All vegetarians said they had heard being omnivorous is part of the human nature. In this sense, vegetarians act against their own nature. By proposing respect to animals, vegetarians are accused of not respecting their own species and disregard all human evolution history.

What is not visible in this debate, though it is implicit, is that the man-nature relation is historic and subject to transformations, as Keith Thomas [8] demonstrates in his study on man's relationship to plants and animals in England 1500-1800, a period with substantial changes. According to the author, man used to live in a hostile environment and it would be anachronistic to think of cruelty towards animals in a situation that imposed the need to fight to conquer and control the world.

Nowadays, we can see changes in the way to think about nature, which is a symbolic construction, and there are various alternative proposals to interact with it. One of the conflicts on what the natural world is can be observed in the context of eating. Vegetarianism, in the sense adopted in the testimonies, can be an example.

2.3 Science

The search for coherence finds in science, or in its jargon, an allegation to defend its options. Both vegetarians and the ones against vegetarians appeal scientific allegations. The discussion is, mainly and recurrently, according to testimonies, about the genetic tendency to vegetarian or omnivorous diets, about the risk of lack and excess of proteins. Within this subject, it all comes to health, a highly valued capital in contemporaneity:

I hate it when people tell me how I replace proteins. I do not eat proteins, vitamins, carbo or other scientific names. I eat food. Then they tell me it is not healthy, but I am not worried with my health, I am worried about animals' health. (Man 2)

If rationality is not the only thing that build an eating habit, some scientific allegations, or science-like allegations, despite the popularity they use from media, do not seem to be enough to change someone's diet, even though doctor's prescription cause some effect on people with heart issues, diabetes, obesity, and others, when they follow some specific dietary prescriptions. Omnivorous and vegetarians discuss about jawbone shape, presence or lack of proteins, length of intestine, among other topics involving scientific terms as a resource of persuasion.

In the dispute about which one is the best, vegetarians and omnivorous individuals appeal to religion, nature, and science to defend their consumption, but what really seems to be at stake is the system of cultural relevance.

3. About vegans

In the world today, we watch the interdiction of all types of animals' meat for reasons of beliefs and health, ethical motivations, and environmental concerns. They question or neglect an alleged humanity omnivorous nature, as discussed in the previous topic, and for these reasons, I highlight the abolitionist vegan aspect. As mentioned, the total or partial restraint of meat can be religious, secular, based on ethical principles, nutritional beliefs, which they defend according to some more or less irrational allegations.

Resuming the classification by Beardsworth and Keil [4], there are six general types of vegetarian diets, which are not fixed, so they can transit among them both ways, from the most rigorous to the least rigorous one and vice-versa. In this most radical diet, there are the strict vegetarians, as vegans are called, one of the groups that compose this subchapter.

Eating vegetables exclusively is the main element that compose the vegan diet, but this one is not only restricted to eating.

Despite many reasons for not ingesting meat, what really motivates veganism is the conception that its followers have about non-human animals. In synthesis, they have the conviction that these are animals with sentience, they are able to feel pain and suffering and have their own interests, and there is no distinction between them and humans that justifies their exploration, slavery, torture, and slaughter.

The conception of humanity and animality is not fixed and not natural among times and cultures. Thomas [8], for example, recalls the creation of the Society for the Prevention of Cruelty to Animals (SPCA) in 1824, in England. This society still exists, now with the addition of the distinctive term Royal to its name, a gift from Queen Victoria. In this same country, in 1944, Donald Watson founded a society against animal exploitation: The Vegan Society.

It is important to emphasize the existence of simultaneous values: the man of science of the industrial era, based on Cartesian allegations, justified animal exploitation for they believe they were just automata. However, Thomas [8] highlights the affection that animals start to enjoy and that today we qualify as companion animals or pets. The societies of protection are born in England, in a country where gaming was one of the most refined sports.

Back to the field of science, even though and despite all questions raised by scientists, including Charles Darwin, who lists man in the animal category, the anthropocentric border remains, and only some of them see the approximation with animals, mainly the mammals or those who have conquered some human affection. In this case, scientific allegations – the strongest ones – end up submitted to relations of affection.

Along with various social moments and with the aid of different scientific signatures, the human supremacy goes under a new review and Richard Ryder [9] coins the term speciesism, in 1970, to mean an asymmetry between humans and other animals.

In this point of view, the so-called speciesism is exactly analogous to racism and sexism, that also describe man supremacy (qualified as masculine, white,

European, and western) above all the rest. Nevertheless, this term raised controversies among antiracist and feminist movements about the analogy experienced by the explored ones. After all, is talking about holocaust of cows in slaughterhouses the same as talking about The Jewish Holocaust? Is the slavery imposed to Africans lived similarly as the confinement of animals? Can milking cows and factory-farm chickens be compared to sexist relations lived by women?

For some, the analogy between speciesism and other liberation movements is viable and enriches everyone in terms of power and voice; for others, the comparison is exaggerated. It seems that feminist ecology has more sympathy to movements related to animal rights, because females are exactly the most explored ones by the industry: for milk, eggs, frequent pregnancies, rape, etc., which draws more empathy in women.

Two authors, among others, appear in the discussion about animal rights. The philosopher Peter Singer [10] claims that animals are sentient and have their own interests and it is not ethical using them for human interests. Tom Regan [11], in the field of law, questions the use of arbitrary weights and measures compared to a distinctive treatment among species, for there would be no difference between the humanity and the rest of the living beings that would rationally justify a pending scale.

The discussion proceeds towards an endless path, walking from the real desire about having a rigorous anti-speciesism attitude; finding yourself deciding whether you are going to take your son to vaccination or not, because it has been tested in animals; giving dog food made with animal by-product to a dog that has been rescued from the street; or willingly or unwillingly killing an ant. The animal rights movement divided itself mainly in these parts: (1) Abolitionists: Contrary to any type of dominance of an animal of any species. Abolitionists believe that all types of interspecific relations will end up being asymmetric and so, instead of abolishing the exploitation. For that, they defend the abolishment of any relation, including with pets and they are, despite the coherence of their speech, impracticable in every single way; (2) Welfarists: They carry the flag of better life conditions for animals, including revivifications in the field of law. For instance, welfarists are favorable to a more human cattle farming, even if it comes to beef cattle. In general, activists that are more radical criticize them and accuse them to defend only the animal welfare for the benefits they can obtain from these practices, like havening more tender meat; (3) Protectors or rescuers: they are not necessarily against beef cattle farming and not against consumption of meat, but they offer temporary or permanent shelter to species elected in terms of affection, specially feral cats and street dogs, but occasionally they also manifest against animals used to pull wagons, tortoises, guinea pigs, and others that, for a given reason, moves someone.

We have generally described the broad terms of these animal right movements, each one their own way, that fight for animals, which are incapable of fighting for their own cause.

At this point, we understand some subdivisions of the movement, because there are people who claim to be vegans, and have a restrict action to the boycott of meat consumption or clothing industry, and are less rigorous about animal products in general. Others get more directly involved in political causes; they free animals from captivity; and actively intervene for changes in the law related to animal rights, among other manifestations. However, in the case of consumption and lifestyle, the fundamental terms of this discussion, we can affirm that an ideal-typical vegan: (a) Refuses to ingest animals and animal products like meat, eggs, milk, honey, and gelatin desserts. (b) Refuse to consume clothes, accessories, and shoes made of animal products. (c) Refuse to consume health, hygiene, and esthetic products tested on animals. (d) Oppose to vivisection as a pedagogical practice at universities. (e) Oppose to the use of animals in scientific researches. (f) Oppose to entertainments that use animals, like rodeos, circuses.

Although the outlines of what a vegan "must be" are clear, the everyday life comes with surprises, from food for your pet, taking your kid to vaccination, being or not being a new target of companies that produce animal products. After all, would it be illicit for a vegan to consume a vegetable burger produced by a famous company that makes lots of profit producing other burgers made from the slaughter of cattle and chicken? The companies are interested in catching vegans, but that is when they should watch what they say. Deciding what to put in a shopping cart becomes an ethical dilemma; deciding whether adopting or buying a pet; offering or not offering vegetables to carnivorous pets in apartments; demanding or not demanding children to follow a vegan diet and restrict their socialization in children's parties. At last, the idea is coherent and of easy understanding, but the difficulties to practice it vigorously are much more difficult and what one lives with it, at the end, is the biggest possible coherence in a world of incoherence and inequality.

4. Conclusions

The results of this research show that the option for vegetarianism or veganism finds resistance and it is subject to everyday embarrassments; nevertheless, despite the divisive role played by vegetarians and vegans in rituals surrounding eating, sociality prevails. Through the data, we realize that the negotiation, the refusal, and the acceptance of varied diets help understand the complexity this decentered society today, which favors the dilemma of individual choices, elaborated by available information and social life embarrassments, whose patterns are fragile. In the intertwining of these vectors, the options for consumption as well as the refusal of consumption provide social roles, communicate social places, and favor the reflection about the contemporary society and its multiplicity.

Thanks

My thanks to Master Fernando Gil for the translation into the Inlesa language and final revision of the text.

References

[1] Douglas, M. Isherwood, B C, The World of Goods. 2nd Edition, Routledge; 1996. ISBN: 978-0415130479

[2] Featherstone, M, Consumer Culture and Postmodernism, 2nd Edition, London: Sage Publication; 2007. ISBN-13: 978-1412910149

[3] Fox, N. Ward, K, Health: Ethics and Environment: a Qualitative Study of Vegetarian Motivations. Appetite; 2008. ISSN: 0195-6663

[4] Beardsworth, A. Keil, T, The Vegetarian Option: Varieties, Conversion, Motives and Careers, The sociological Review, 40; 1992. Online ISSN: 1467-954X

[5] Whorton, J C, Historical Development of Vegetarianism, The American Journal of Clinical Nutrition, 59: 1103-1109. ISSN: 0002-9165

[6] Lipovetsky, Gilles, The Paradoxical Happiness – Essay on Hyperconsumption Society, Companhia das Letras, São Paulo: 2007 ISBN: 9788535910933

[7] Mauss, Marcel, *Essai sur le don. Forme et raison de l'échange dans les sociétés archaïques*, Paris, PUF; 2007. ISBN: 978-2-13-060670-3

[8] Thomas, K, Man and the Natural World: Changing Attitudes in England 1500-1800, England, Front Cover, Penguin; 1984. ISBN: 0140073442

[9] Ryder, Richard D, Speciesism, Painism and Happiness: a Morality for the Twenty-first Century, Societas, United Kingdom; 2011. ISBN-13: 978-1845402358

[10] Singer, P, Animal Liberation, Review of Books, Harvard; 2001 ISBN-13: 978-0060011574

[11] Regan, T, Empty Cages: Facing the Challenge of Animal Rights, Lanham, MD, United States; 2005 ISBN-13: 978-0742549937

Chapter 6

Plant Proteins as Healthy, Sustainable and Integrative Meat Alternates

Abstract

Vegetarian protein diet based food industry have emerged as one of the fastest growing industries with largest than ever shelf space it has created in today's market. The rapid growth of plant protein industry is attributed to increased health awareness, economic and environmental sustainability issues of animal proteins and their nutritious, economical, and healthy food image among masses. Technological interventions like extrusion texturization has enabled the food engineers to create the imitation meat which approximates the esthetic attributes (texture, flavor, and appearance, binding ability, chewiness, firmness or softness) and/or chemical nature of meat. These texturized plant proteins are healthier and economical meat substitutes with sufficient opportunity to manage modify or change their functional properties in accordance to specific consumer demands.

Keywords: meat proteins, protein transition, sustainability, plant proteins, innovation, plant-based meat alternatives

1. Introduction

Meat on behalf of satisfying all the basic urges of consumer as a typical textured, juicy, flavorful, chewy food is the non-vegetarians first preference which also fulfills all the nutritional requirements of the consumers too. But the negative impression tagged with meat from very beginning always questioned its consumption may be it is related with ritualistic aspects or their significant effect on the environment due to an inefficient consumption of energy and land along with the emission of gases by meat production units [1]. The organizations and the policy makers which are involved with the sustainable production and consumption are expecting to make a shift towards more sustainable products for the consumers. However, certain products can be tagged as an alternative to the market meat which covers very less proportion of the market, called as meat replacers or meat substitutes. To overcome the dilemma of entire situation for the production of sustainable food to fulfill the needs and requirements of meat lovers the food researchers and processors formulated "meat analogue" providing higher or at least same nutritional and health benefits along with the satisfaction of meat consumption. Analogues are the structurally similar compounds slightly differs from one another on the basis of composition, in the case of meat, analogue are the products structurally similar to meat but differs

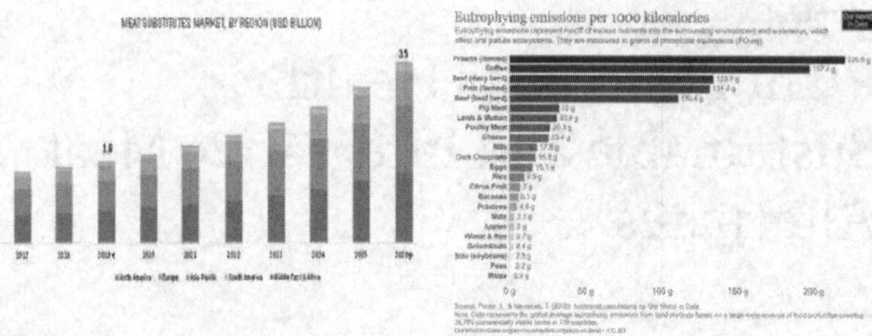

Figure 1.
Growth of meat substitutes in different continents and the eutrophying emissions of various food commodities.

in composition. They are also known as mock meat, imitation meat or faux meat approaches the qualities (mainly flavor, texture and appearance) esthetically present in meat and also fulfills the chemical characteristic of particular meat type. Such as "surimi" which is a meat-based, less expensive/healthier alternative particularly to a meat product. A general meaning of meat analogue is the food constituting non-meat ingredients sometimes without any inclusion of dairy products however, fulfilling all the nutritional and characteristic requirement of actual meat product. The meat analogue covers the maximum market which includes either vegans, vegetarians and the non-vegetarians who are in urge of reduction of meat consumption either for ethical or for health reasons and also includes people with dietary laws based on religion, such as Buddhist, Halal and Kashrut which is expected to be increasing in the upcoming years according to the data given in **Figure 1** globally the meat substitutes market was assessed to account for USD 1.6 billion in 2019 and is projected to reach USD 3.5 billion by 2026, recording a CAGR of 12.0% during the forecast period. The market is primarily driven by the increasing demand for plant-based meat products among the millennials.

The base of some vegetarian meat analogues is built on the recipes which are older than centuries viz., rice, wheat gluten, legumes, pressed tofu or tempeh with flavor addition which make them to taste like beef, chicken, ham, lamb, seafood, sausage, etc. Another meat analogue based on soybean constituting boiled soy milk layered by a thin skin is called as Yuba. Some more recent developments in soy based meat replacers is TVP (Textured Soy Proteins) derived from dry bulk commodity from soy, myco-protein-based quorn excludes vegans because it contains egg white which acts as a binder. Meat analogue functions as a meat replacer in the diet. The market does not only include vegetarian population but also include non-vegetarians who are in the urge of reducing their meat consumption either due to ethical or health perspectives. These innovations are the advantageous to the people who are concerned about the health related problems arises with the fat, cholesterol and salt overconsumption. There is also a need to develop new ways for the fulfillment of the nutritional requirements of poor at a minimum cost. Thousands of plant proteins are fortunately available in the world, explored or yet to be explored for the production of meat alternatives.

2. Driving forces for the development of meat analogue

Although the meat has a rich nutrient composition which includes essential nutrients like proteins, micronutrients such as zinc, iron and vitamin B_{12}, but taking sufficient intake of these nutrients is possible without taking meat in the diet with the consumption of variety of other foods available. In western countries, where

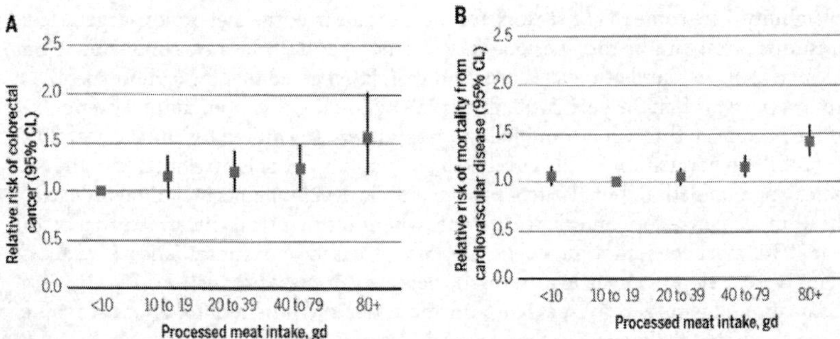

Figure 2.
Incidence of colorectal cancer (A) and development of cardiovascular diseases associated with red meat consumption.

various meta-analyses and prospective studies had undergone shows a moderately high mortality rate than usual in participants consuming more amount of processed and red meat. High intake of processed meat increases the risk of colorectal cancer is the strongest evidence of the adverse effects. The relative risk of colorectal cancer [2] and cardiovascular deaths [3] as a function of average processed meat intake shown in **Figure 2.** The International Agency for Research on Cancer (IARC) under World Health Organization categorized red meat and processed meat, due to its association with colorectal cancer probably being carcinogenic to human being [4]. According to an estimation by IARC 34,000 deaths due to cancer worldwide per year are attributable to diet high in processed meats and if the associated reports with the eating patterns of red meat were taken casual, 50,000 cancer deaths could be possible worldwide per year due to the high intake of red meat [5]. In Western Europe, the processed meat average intake is 26.4 g/day [6] will lead to an increase in the risk of colorectal cancer upto 9%. As compared to plant based foods meat produces more per unit energy emissions because there is a loss of energy at each trophic level. Within the types, the energy emission increases with poultry, mammals and ruminants production. As a reason for methane production, meat is a single most important source, which is known for its high warming potential, significantly low half-life as compared to CO_2 in the environment [7]. Grassland system careful management contributes to the storage of carbon, but it cannot be considered as a profitable effort. The use of fresh water in the agriculture field is anyhow more than any other human activity, in which the one third portions is required for the livestock, so in the areas where there is less water availability meat production is a major competitor with the other water use which also includes the maintenance of natural eco system. Production of meat is an important source of phosphorous, nitrogen and the cause of other pollutants that affects the biodiversity particularly through conversion of land to pasture and arable feed crops. **Figure 2** represents the nutrition runoff into the surrounding environment and waterways which pollute ecosystem, where meat and meat products shows maximum pollution rate.

3. Comparative overview: plat proteins vs. animal proteins

3.1 Economy and sustainability

From our ancestral period onwards, meat plays a distinct role among other food sources. Comparatively high protein content, energy that are readily available, palatability – are some major reasons for meat to rule over the diet of non-vegetarian consumers from pre-historic period. Various studies and surveys are conducted to note the increasing demand of meat in near future. It is expected that for 6 billion people around the world, in the year 2000, about 229 billion Kg of various animal products where produced; this will be doubled (approximately 465 billion Kg) by the year 2050 for 9.1 billion population [8, 9]. This hike in demand is not only attributable to the increasing population, but also to global economic developments including industrialization, urbanization, and rise in income, where another study illustrates that by the year 2030, meat consumption will be 72% more than the current statistics. Livestock farms were used previously to achieve the necessity of meat production [10, 11]. Today, limited natural resources like land and water and other ethical issues are some of the inevitable challenges for the sustainability of livestock farms.

In present scenario, about 20% electricity from total energy generated and 70% of water from total freshwater consumption is utilized for the production of food crops. Majority of thus produced crops and its wastes are used for the growth of livestocks in the form of feed and fodder, i.e., converting plant protein to animal protein for human consumption. Briefly, the reality is, when the world is facing food security and scarcity of natural resources, we are wasting majority valuable resources for the conversion plant nutrients to animal products. For instance, a study states that out of total grain and soy produced, 40% of grain and 75% of soy is used as feed. Then again, ideology of people matters a lot in order to put an end to these disputes. Studies conducted in various parts of the world give a common sketch that mostly, consumers who are health concerned is rarely diverting to the meat substitutes, and surveys shows that female population and educated people are more interested to substitutes animal protein than male population and less-educated counterparts. Factors such as awareness of the negative health impacts of meat products, comparatively lowering number of neophobic people (who shows an extreme or irrational dislike towards anything unfamiliar or new to them), transition to plant-based diet, and publics who endorse to protect and care nature are some of the reasons for the growth and sustainability of meat substitutes; indirectly through which economy and sustainable development of a nation is influenced [12].

3.2 Nutritional and functional properties

Consumption of plant and animal products always depends upon many reasons. One among them is their nutritional profile. From macro-nutrients like carbohydrates, protein and fat to micro-nutrients like vitamins and minerals varies drastically in both the sources. Due to the reason of palatability, it is commonly observed that consumption rate of animal products, including dairy and egg are always higher than the plant products. Concepts of higher biological value of egg, whey, etc. than the proteins of soy and other plant sources are also an inevitable factor for increase in the consumption of animal or meat products.

From past decades, researches have provided sufficient evidences to increasing health risksdue to meat consumption.Animal proteins are usually associated with SFAs and plant proteins with fiber and phenolic compounds. This is usually reffered as "protein package" or "whole food package". Studies conducted among no-vegetarians, pesco-vegetarians, and vegetarians show that meat consumers are observed with higher energy intake along with high quantity of SFAs, vitamin B_{12}, vitamin D, zinc, and iodine, while fiber, PUFAs, vitamin C, vitamin E are in lower concentration. Higher intake of nutrients and components like heme, cholesterol, etc. from animal sources, especially red meat, can cause to cancer, type-2 diabetes, stroke, heart disease, and metabolic syndromes. On the other hand, plant protein, regardless of the sources, are consistently related to nutrient adequacy.

Studies on substituting animal protein with plant protein are conducted widely and some of the results include: about 21% of diabetes risk was minimized on substituting processed meat, where dairy product showed no great change [13]; red meat and other high glycemic foods such as refined grains, potatoes, etc. account for high risks; purified protein (e.g. soy protein isolates) ingredients on higher consumption (~35 g/d) has a negative impact on blood lipids and blood pressure, while lower dose of soy protein (say, < 25 g/d) shows no remarkable effect [14]. Thus, both animal and plant proteins has their own influence over the consumers, where quantity and type of nutrient are important.

However, predictably, plant proteins generally show some health benefits especially against cardio-metabolic risks and other chronic diseases. This is possible by replacing many components associated with animal protein like saturated fatty acids with plant protein package. Also, inclusion of higher quantity of plant source helps in consuming fiber which has prebiotic effects, thus assisting the beneficial micro flora in gastro-intestinal tracts. Additionally, amino acids also has influence on various diseases to great extent. For example, arginine acts as a vector in vascular homeostasis. Various studies concluded that non-indispensableamino acids (arginine, glycine, cysteine andglutamine/glutamate) with potential benefits are mainly seen in plant proteins.

4. Production technology

Invasion of plant-based substitutes in our daily diet has been started decades before, whether it is milk analog or meat analog. Many of the foods we consume today consists of plant protein and had been used as a source of plant-based protein long-ago to substitute and reduce the consumption rate of meat and meat-derived products. Some of the traditional recipes include mushrooms, legumes, tempeh, wheat gluten, and pressed tofu in which slight alterations are made sometimes by adding flavors to obtain final product that can successfully imitate the meat (lamb, chicken, beef, sausage, ham, etc.) in sensory attributes. Enlightenment on the health benefits of plant-based protein over the various adverse effect of meat consumption caused a major impact on increased production rate of meat substitutes in the food market around the world. Thus many innovative foods that can replace meat in table are available easily. Generally, the production of meat substitutes in supply chain is broadly classified into four major steps: [15, 16].

i. Protein crops are identified with better quality and cultivated globally,

ii. Crops are undergone various processing methods to obtain protein ingredients such as protein isolates and concentrates,

iii. These protein ingredients are then formulated and processed into texturized intermediary products and final meat substitutes are developed,

iv. Products are distributed among consumers through retail and different food service method.

Various processing technologies are utilized for the conversion of plant-based protein into a product that can imitate meat especially in the area of texture and taste. Conceptualization of technological development in food sector is differentiated as First generation meat substitutes (mainly composed of low-moisture cooking developing intermediate products) and Second generation meat substitutes (based on high-moisture cooking extrusion, started in European markets at early period of 2000s). First generation meat substitute was found to be existing since 1990s in European markets and includes product like Textured or Texturized Vegetable Protein (TVP). The method is based on the working principle of extrusion consisting screw system within a barrel. In this method, high temperature is provided to the raw material and compressed inside the barrel, which is then conveyed through dye(s) so that an expanded final structure is obtained. This undergoes further processing to form final meat substitute. On the other hand, second generation meat substitutes were a success due to the advancement in food sectors, especially in the areas of cooking extrusion technology, various innovations in hydrocolloids and so on. These technologies opened a new insight to food market, not just because they helped in developing products that are more appealing and similar to meat, but also application of wide range of raw materials categorized as cereals, pulses, legumes, oilseeds, and aquatic plants. **Table 1** briefly explains some commonly used protein sources and its corresponding market products along with health benefits and challenges. Based on the availability, cost, convenience, nutritional profile of both raw material and final product, functional characteristics and physiological properties, numerous meat analogues are developed and are still in its experimental stages [25]. Acceptance of the final product of a meat substitute, however, has a vast influence on the ingredients used while manufacturing. **Figure 3** gives an idea on general classification on the ingredients for developing a meat analog.

Primary ingredients or protein sources	Meat substitute products	Health benefits	Challenges	References
Soy bean	Tofu, tempeh, Texturized soy protein (TSP) Schnitzel	• Reduce of chronic illnesses (CVD, immune disorders, diabetes, obesity, anti-hypersensitive andbreast and prostate cancer • Nutraceutical effects • Anti-fibrosis, anti-estrogen • Shows positive effect on thyroid and fertility	• Development of hormone-responsive tissues are reported • Beany flavor reduces the consumer acceptance	[17, 18]
Wheat	Seitan	• Helps in bowel movement • Increase gut microbiota • Decrease of serum cholesterol levels • Reduce post-prandial blood glucose level • Lowers the risk of cancer and colitis	• Gluten is a poor source of essential lysine and threonine • Many of wheat proteins are recognized as allergens and triggers of celiac disease (CD)	[19, 20]

Primary ingredients or protein sources	Meat substitute products	Health benefits	Challenges	References
Legumes (pea, lentil, lupin, chickpea)	Lupin steak and other similar products	• Gastrointestinal and gut health • Antioxidant, anti-inflamma-tory, anti-cancerous and cardio-protective activities • Obesity and weight management	• Presence of antinutrients and some toxic com-pounds like phytic acids,lectins, saponins,alkaloids, tannins and enzyme inhibitors decreases protein digestibility and bioavailability of minerals	[20]
Oilseeds	CanolaPRO™	• Reduce plasma cholesterol level by inhibiting intestinal absorp-tion cholesterol • Decrease cell proliferation • Anti-inflammatory, antioxidative, and anticarcino-genic activities	-	[20]
Filamentous fungus *Fusarium venenatum*	Mycoproteins Quorn™ products (e.g. Meat free sausages, Meat-free chicken and Apple sausages)	• Low fat content and high protein content helps in lowering blood cholesterol level, and has glycemic properties	• Increase of uric acid level in blood leads to urolithiasis • Mycotoxins and allergies in some cases are still a challenge • Acceptance on the basis of sensory attributes like taste and color	[21, 22]
Algae *Arthrospira platensis* (spirulina)	Sausages	• Pigments like Fucoxanthin present in brown algae helps in apoptosis in human cancer cell line • anti-inflammatory, Antidiabetic, antioxidant properties • Eicosapentaenoic (EPA) &doco-sahexaenoic acids (DHA) are against chronic diseases	• Consumer response to green protein products due to chlorophyll • Requires complex and effective pro-cessing methods • Waste water gener-ated may cause hazards like algae parks	[23, 24]

Table 1.
Table presenting different protein sources, corresponding meat substitutes and challenges faced for its acceptance.

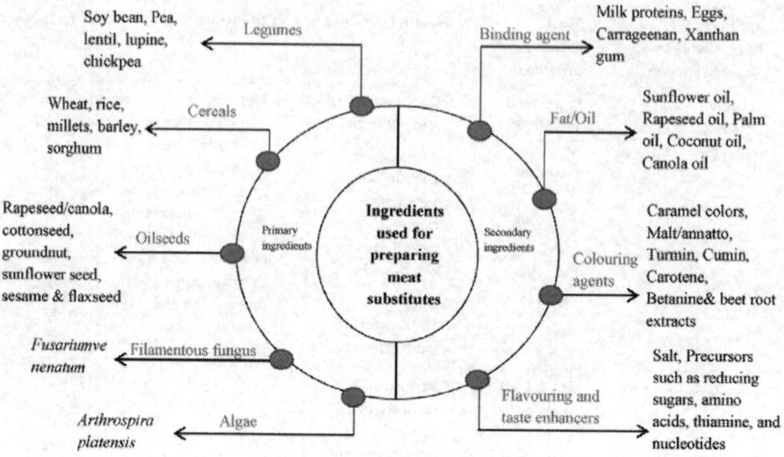

Figure 3.
Schematic representation of ingredients used in meat analog industry.

5. Common meat – substitutes available in the market

5.1 Soy-based products

Soy is considered as one of the raw material used traditionally for the preparation of various meat substitutes. From soy flour to soy protein concentrate and isolate, they have acquired quiet a significant position among the recipes. Among different soy proteins ingredients, its isolates are not only highly pure, but also has a light color with bland flavor, which makes them more approachable in product development, even though degree of purity does not play much role in the meat-analog applications. Properties like water-holding capacity, gelling property, fat-absorbing capacity, emulsifying capacity and other functional properties of soy ingredients makes them more reliable in this industry and are available in numerous forms like toasted flour, full-fat, de-fatted (about 50% of protein content and are produced from grinding defatted soy flakes), etc. By fractionating defatted soy flakes, soy protein concentrates (protein content is found to be 70%) and isolates (protein content is 90%) are obtained through aqueous alcohol extraction and alkaline extraction followed by precipitation in acidic pH, respectively [26].

5.2 Tofu

Soybean derived tofu or soy curd, being an excellent source of protein and minerals like iron and calcium, is the mostly utilized meat substitute world-wide and are available in block form. Production of tofu is said to be simple, clean, convenient and controllable process. Nutritional profile of tofu includes 8% of total proteins, 2% of carbohydrates and lipids about 4 to 5% on fresh weight basis. Absence of cholesterol, low energy value, high amount of vitamins and minerals and presence of dietary fibers (about 1%) are some of the relevant factors for the high demand of soy curd. Traditionally, tofu is prepared by protein coagulation of hot soy milk with the assistance of a coagulant: salt-induced($CaSO_4$, $CaCl_2$or $MgCl_2$) or acid-induced

(Glucono Delta-Lactone (GDL) – also known as gluconolactone) to obtain a gel-like product. Some studies says that coagulant used for the production of tofu plays a vital role in determining its quality; yet controversies are still existing.

The complex process starts from soaking and grinding of soybeans followed by filtering to extract the milk. Then it undergoes methods like boiling, coagulation, breaking of curd, pressing and reforming the gel. However, packed or filled tofu involves addition of coagulant to the cooled soy milk and is then heated and coagulated in a suitable package, without undergoing any further processing methods. Some studies says that, presence of isoflavones, oligosaccharides, trypsin inhibitors in soymilk cause allergic reactions, off-flavor and flatulence, respectively in some people. As a result, soy protein isolate (SPI) has gained a considerable demand as raw material for packed tofu, as it has better health benefits to consumers. Apart from coagulant used, some other influential factors in tofu-making are the processing conditions, concentration of coagulant used, and two major storage proteins: glycinin (11S) and β-conglycinin (7S) [27].

5.3 Tempeh

Tempeh, a traditional soy fermented product, is considered as the food that can provide the most health benefits among other soy products and higher consumption percentage in many places. Countries like Indonesia (70% of households), Australia, China, Japan, and Taiwan and also in some places of Europe, America and Africa are reported to have higher consumption of tempeh. Presence of high quantity of crude protein and essential amino acids, fatty acids, carbohydrates, folic acid (416.4 μg/100 mg), isoflavones, vitamin B_{12} (3.9 μg/100 g), and tocopherols (α-, β-, γ- and δ-), along with reduction of antinutritional factors such asphytates, saponins, trypsin inhibitors, hemagglutinins, and flatus factors, with increased facilitating, detoxification properties, bioavailability of minerals and many otherhealth benefits, tempeh is considered as a better choice for healthy diet [28]. Presence of umami taste (a basic taste that can be detected by human receptors) is also said to be a factor for its increased demand.

Tempeh, in general, is a collective name for combination of cooked and fermented raw material (cereals, beans or a byproduct of food processing) with any suitable culture ofmycelium of a living molds, yeasts, lactic acid bacteria (LAB) or various gram-negative bacteria. Even though *Rhizopus oligosporus*is the dominant microorganism used for the production of tempeh, molds like *R. oryzae* and *Mucorspp* are also used for enhancing flavor, texture and nutritional characteristics.

Production of tempeh involves acidification of soybeans using lactic acid or acetic acid to inhibit the growth of undesired microorganisms by lowering the pH (5 or below), followed by boiling and cooling (30–38°C). Inoculation at 25–30°C process is carried out and a compact, creamy, white, fresh tempeh cake will be resulted after 1–2 days. Due to the production of different proteases secreted by *R. microsporus var. oligosporus*such as 'aspartic acid protease' or known as 'acidic protease' and another endopeptidase called 'serine protease' helps to break complex soy proteins at aspartic acid residues (when at 3–4 pH) and at small/side chains like glycine and alanine residue (at neutral to alkaline pH, i.e. 7–11), respectively. However, for commercial purpose, mixed cultures are also used for better and quality yield [29]. In addition, fermentation process not only helps to improvise the nutritional and sensory profile of the final product, but also imparts health benefits including enhanced antioxidant property, and plays a role in fighting dementia, cardiovascular diseases andcancer (especially, colorectal cancer and hormone-depended cancers like breast cancer and prostate cancer) [28].

5.4 Textured soy protein(TSP)/soy meat

In food industries, inorder to reduce the rate of saturated fat and cholesterol consumption, vegetable proteins (VP) are incorporated into food, increasing the protein and essential amino acid content. Sources of all vegetable proteins like legumes (protein content is maximum, varying from 25 to 50%), nuts and soy are used for the production of Textured Vegetable Protein (TVP), which is found in fibrous, insoluble and porous form and considered as an excellent meat analog or meat substitute. TVP is otherwise known as textured soy protein (TSP) or soy meat, where proteins present in defatted soy flour is concentrated, isolated and extruded based on 'extrusion cooking', providing better taste and chewy-texture when compared to meat or seafood.Soy meat has about 50% protein content, which is found to be decreasing on rehydration.This method was primarily adopted by many Asian countries. TVP was commercially developed in America and by late 1960s, it was successfully welcomed by European markets.

Soy meat is presently considered as an economic option for replacing meat (e.g. meatballs) by both vegetarians and health conscious consumers. TSP is used as a meat extender in many products, thus replacing 30% of meat without affecting the sensory characteristics. For instance, in some parts of the world, quantity of beef in samosa stuffing has been replaced around 50% with granules from defatted soy flour with no major observable difference in sensory attributes. Production of TVP starts from washing of the selected soybeans and is soaked at 30°C for 3 hrs in order to remove antinutritional factors by softening the husk. Soaked beans are then washed till the husks are completely removed and are dried at 70°C for 5 to 8 hrs. TSP is developed through 'hot extrusion' where different dyes are used for producing high protein nuggets, chunks, etc. Today, soy meat is expanding worldwide rapidly, especially in the developed countries. Since soy meat is considered as "poor man's food" by many, it is a reliable source of protein for under-developed countries and low-income people [30].

5.5 Cereal-based products

Cereals comprise of nine species (wheat, rice, corn, barley, sorghum, millet, oat, rye, and triticale) under the family *Gramineae.* In meat analog industry, cereals are mainly used for extending the meat products. For example, 3–9% of quail meat roll is extended with corn flour resulting better emulsifying stability with yield. Similarly, chicken patties are also substituted with 10% barley flour, 5% sorghum and 5% pressed rice, which has no significant change in its sensory profile. Studies with rice and barnyard millet includes the decreased level of cholesterol and increased magnesium content among the consumers and use of these cereals has also not shown any negative impression on color, except barnyard millet slightly affect the flavor of developed meat substitute. Today industries are substituting cereals 9% or less, which is more convenient. Thus, cereals play a vital role as an important ingredient, particularly as a meat extender. In some cases, combination of cereals gives better yield and quality to the product, without adversely affecting its sensory. On the other hand, presence of gluten in cereals like wheat, oats, barley and rye arise question among the consumers due to its potential of allergy [31].

5.6 Seitan

Wheat protein has been playing a vital role in human diet from thousands of years ago. Traditionally, they were used widely in the regions of Japan, Korea, China and Russia, to replace meat products. Wheat being a common cereal used in majority of countries, it is possible to rely on wheat-based meat substitutes, especially in regional level. "Wheat meat" or "wheat gluten" is developed from the component called gluten (wheat proteins -gliadin and glutenin), which is easily extracted through a simple procedure of rinsing with water in order to remove starch and bran. As a result, a chewy mass is obtained, which can be further processed with additives and cooking methods to attain wheat-based meat substitute. This simple, conventional and economic method is utilized by many food industries that deal with vegetarian burgers, sausages, minced meat, nuggets and schnitzel. Mostly, rinsed out starch is utilized as byproduct for other purposes, thus reducing the food waste. When gluten is flavored by simmering in a broth of soy sauce, garlic and ginger to obtain seitan, which has small quantity of sodium. Gluten has the capacity to form a thin film on elongation resulting a natural stringy fibrous proteinaceous structure seitan, which helps it to imitate the texture and consistency remarkably. Seitan is not just an alternative for the non-vegetarian diet, but also it is an ideal choice for people who are reluctant to consume soy products. Even though many nations like Western countries started to use seitan as a part of their food, some Arab countries are not in favorable due to its unpleasant flavor, failing the sensory attributes [32].

5.7 Mycoproteins

Apart from plant-based proteins, scientists diverted their experiments towards single-cell proteins (SCP) for the development of meat substitutes. SCP can be described as protein derived from pure or mixed cultures of microbes such as bacteria, yeast, fungi or microalgae. Most common source of single-cell protein is the filamentous fungi, which produce mycoproteins when grown under specific controlled environments inside a bioreactor. Processes like forming, steaming and subsequently texturizing are done to achieve finished products. Sometimes, for obtaining desired products, binding agents like egg albumin and flavoring agents are also used accordingly. Mycoprotein is also known as 'fungal protein' and described as "Generally Recognized as Safe". Commercially, *Fusarium (F.) venenatum* is widely used in food industries for the production of mycoproteins and industrially known as Quorn™.

Studies on fungal mycelium for substituting meat were started during 1960s by Rank Hovis McDougall (RHM), a British company. After fruitfully completing the development of product and its toxicity testing, first product was launched in the year 1985 with the approval from British Ministry of Agriculture. Thereafter, many researches are revolving around mycoproteins. For instance, a study on the biological value of mycoproteins was found similar to the milk proteins and toxicology study says that these fungi derived proteins have no harmful effects on human beings and animals [33]. Additionally, it provides some health benefits to the consumers suffering from various chronic diseases. Presence of protein (45%), carbohydrates (10%), fat (13%), fibers (25%) on dry basis along with various vitamins and minerals helps them to replace several meat products in our food basket. Also, studies have proved that intake of mycoproteins have a positive effect on lowering total and LDL (low-density lipoprotein) cholesterol and increasing HDL (high-density lipoprotein) cholesterol level in blood; appetite regulation as short-chain fatty acids (SCFAs) produced as a result of gut digestion of mycoprotein fibers send satiation signals from colon to brain; presence of soluble fiber not only helps to slow down the diffusion of glucose molecules through the intestinal walls, but also decreases the rate of absorption of glucose, thereby having a positive outcome on glycemic condition; and they can fight against food-borne pathogens which are commonly seen in numerous meat products [21].

Marlow Foods, developer of Quorn, develops mycoprotein through continuous fermentation of desired culture accompanied by glucose and other required nutrients and oxygenated water. In US market, about fifteen Quorn products are sold, among which products: Meat-free sausages, Meat-free chicken and Apple sausages are refrigerated items [23]. Production method of mycoprotein is said to have a favorable influence on environmental issues also. That is, they can reduce the amount of greenhouse gas emission from agriculture and food allied sectors, use of innumerous fertilizers, antibiotics and pesticides can be decreased to a large extend, wastage of land and water and nutrient-cycle recycling, especially reactive nitrogen species (RNS) - NH_4^+, NO_3^- and NO_2^- by fixing nitrogen gas. However, many regulations are adopted by many nations due to some negative impressions of mycoproteins. Firstly, presence of higher concentration of nucleic acid (NA) in mycoproteins can cause urolithiasis due to increase of uric acid level in blood. Secondly, Use of microalgae and yeast can affect sensory attributes like flavor and color of final product. Thirdly, verifications regarding mycotoxins are inevitable as a part of safety precautions. Lastly, reports on gastrointestinal tract (GIT) reactions, sometimes leads to life-threatening allergies (anaphylaxis and urticarial) [21].

6. Miscellaneous ingredients and their role

6.1 Binding agents

To develop a product that can mimic another is always a challenge. Especially to satisfy its sensory appeal. Additives that are capable of binding water and fat are important while developing a plant-based meat substitute. These ingredients can either be the plant source itself such as isolate or concentrate of soy protein and wheat gluten or some external components like egg, xanthan gum, milk protein or carrageenan. These ingredients, which are having high protein content, majorly functions as water binder and forms protein networks. On the other hand, some ingredients with no protein content (e.g. starch, soy flour, etc.) are used as fillers taking into consideration of their binding properties by entrapping water and fat physically.

However, to attain a quality product, concentration of binding agent used at the time of manufacturing of meat analog is vital. Industries make use of gluten due to its cohesive nature, leavening ability, dough forming ability and visco-elastic nature. In addition to these properties, gluten is profitable while processing because of the reduced cooking loss and can improve slicing attributes [34, 35]. In some case, polysaccharides (e.g. guar gum, pectin, cellulose, carrageenan, etc.) are used as binders. The gelling and thickening properties helps polysaccharides to improve the rheological properties of developed product. Hydrocolloids such as xanthan gum, starches, pectin and locust bean gum are found to have a positive effect on sausages and other similar products which are low in fat content [36]. Other than the application of egg albumin, soy ingredients (flour, concentrates and isolates) are also extensively used. However, beany taste of soy products makes the industries to limit their application to isolates [26].

6.2 Fat/oil

One of the prime objectives of developing meat analog is to reduce the cholesterol level to minimum. As a result, today, meat-substitute products available in the market contain comparatively low fat. Another concern is that, quantity of fat can affect fiber structure and its formation during processing undesirably. Studies reported similar cases, for instance, during extrusion process, dough that have oil content more than 15% was failure as the lubrication of the material affected the alignment of its macromolecules; and also the slippery texture negatively affect the shear force exerted at the time of extrusion [37, 38]. Although raw materials are fatted before using them for manufacturing meat substitutes, plant sources like soy, oilseeds, etc. has natural oil present in them. This quantity is more or less sufficient in many cases. However, for increasing the sensory attributes, industries use some additives such as soy oil, sunflower oil, rapeseed oil, canola oil, palm oil, corn oil, coconut oil, etc. These added oils give the final plant-based meat analogues a juicy and tender texture, along with release of flavor and retaining the volatile components [26].

6.3 Flavoring agents and taste enhancers

Flavor is one of the most important sensory attribute for food as it gives a satiety before even consumption. And in the case of meat products, flavors arising from the product itself make consumers to stick around the non-vegetarian foods. Therefore, while preparing a meat analog, ensuring the customers with similar flavor in essential. In many products, fat/oil is added extra, which can also serve as a precursor for flavoring by entrapping the organoleptic volatile components. Addition of flavoring agents is hence common in meat analog industry. These agents improve the acceptability of product not only by mimicking the flavors of meat products, but also by lowering the beany flavors generated (e.g. beany flavor developed for soy-based meat substitutes). A study conducted on mushroom concentrates for replacing monosodium glutamate or hydrolysed vegetable protein was effective when formulated at less than 1%. Also, simple sugars and sulfur-containing amino acids have strong impact on developing flavor, while other compounds like glycoprotein, monosodium glutamate, etc. helps to mask pungent sulphury taste or improves the meaty flavor [10, 11].

However, during extrusion process (or while undergoing treatments at high temperature and pressure), it undergoes various physiochemical changes, making the whole process a complex. This results in loss of natural and added flavors like spices and other agents. Conversely, application of high temperature give rise to reactions like Maillard reaction, where amino acid and sugars present in the matter creates a distinctive flavor. This improves sensory characteristics – flavor and taste, even though it is risky if not optimized properly by reason of the generation of off-flavor. Among numerous aromas available in meat products, roast and smoked aromas are desired by many consumers. Additionally, several studies shown that furans and thiophenes containing sulfur or sulfur-containing heterocyclic compounds imparts strong meaty strong meaty-roasty-flavor to finished products [26].

6.4 Coloring agents

In meat products, degree of color change and color characteristic are very important, hence it is essential to maintain same color attributes in meat analogues. This can be achieved by incorporating edible additives which imparts desired color in final product. For example, protein from soy and wheat (gluten) possess yellow-brow color and is much brighter than original cooked meat and very different from raw meat color. Edible food color are used such caramel, annatto, carotene, turmin and cumin as they are heat stable. Other colors may be used in meat analogue are beet root and betanine extracts and reducing sugars for browning characteristics in Millard reaction in presence of protein. Such reducing sugars are mannose, lactose, xylose, arabinose, dextrose and maltose. In general, coloring solutions are added with proteins before extrusion. But still majorly meat analogues have poor color quality due to improper balance of pH between meat alternative products and color solution. This issue can be resolve by addition of appropriate acids alone or in combination like acetic acid, citric acid and lactic acids. Also, color retention agents like hydrated alginate or maltodextrin can be added to control the color migration from meat alternative products [26].

6.5 Advantages and disadvantages

The expanding world's population facing the problem due to lack of food led to hunger issues along with preserving ecosystem. In 20th century, the drastic nutritional transition was developed and presents livestock as a major available protein source globally. However, consumption of animal products also reported with various serious food safety issues.

Meat and poultry products are popular and consumed all over the world but many ethical, traditional and environment issues forced the production of meat analogues. Plant based many products were designed recently to substitute the conventional meat products. The basic criteria of protein for the meat analogue are essential such as oil and water holding capacity. During the slaughtering of animals, food animals were reared which is a conversion of vegetable protein to meat protein. This process required many resources like water, fodder, feed and so on. Also, crop production required low inputs comparatively. Thus, production of meat reported with inefficient compare to plant based meat analogue. As per Pimentel and Pimentel [39], meat production required more water in comparison with plant crops (approximately 100 times more). The requirement of water in human diets, for meat diets per person in a year was the double than the vegan diets per person in a year. Meat products greatly depend on meat analogue which further vary with consumer acceptability, cost and legal criteria of the country. Along with this, the current living style associated with health diseases encourages the production of plant based meat analogue. Broadly the major concerns with meat products are classified as environmental issues, animal welfare issues and public issues. The expenditure of huge amount of natural resources and emission of greenhouse gases by the meat productions enhances the ecological burden which led to deforestation also. Animal welfare issues involve the unethically torturing and slaughtering process. This sometimes led to forceful and cruel transportation of animals. Public health issues comes in picture when the over consumption of meat reported resulting in development of ischemic heart disease and cause of 1.8 million deaths every year [40]. Larsson and Wolk [41] reported the colorectal cancer risk associated with consumption of red meat (120 g red meat per day or 30 g processed meat/day). *Salmonella, Campylobacter* and *E.coli* pathogen outbreaks reported several times all over the world which is also found in meat products. These major issues raise the food safety and public health concerns and studies. Meat analogue have advantage of producing various shape and size like sheets, cubes, disk etc. with desired color.

In terms of nutrition and sensory, still no plant based meat product was successfully developed which can completely replace the meat products. The plant based meat substitute lacks with the similar mouthfeel due to the hardy and rubbery texture of meat analogues. The other drawback reported for the plant based meat analogue is off-flavors. Aforementioned, sensory attributes of plant based meat analogue still lacking leading to the undesirable flavors in the end product. Due some allergic reactions, not all plants can be used for production of meat analogues which is an important health concern. Major disadvantage for the meat analogue is related to its production process which involves 4 steps starting from cultivation of plant crop followed by protein concentration. Further, formulation and texturizing of meat substitute and marketing add up the various unit operations. In comparison with animal meat, few unit operations are required.

6.6 Laws and regulations

The health importance in protein in diet is scientifically documented and in 1999, FDA (Food and Drug Administration, U.S.A.) also recognized soy protein as blood cholesterol ingredient when consume 25 g every day when included in regular diet. This statement led to utilization of plant proteins in different health and function food products [42]. In 2016, GPA (Green Protein Alliance: multi-stakeholder platform with other partners) in Netherlands, focused on protein consumption balance to 50:50 from 63:37(animal:plant sources) by 2025 [15, 16]. In a study, an American consume 112 g protein (77 g protein from animal source and 35 g plant source) in daily diet and RDA for protein in adult is 56 g from both type of diet (vegan and non-vegan). These data conclude that consumption of protein is twice than RDA [39]. Incorporation of plant based protein in meat analogue related products will be helpful in balancing the proteins by following standardizing the nutrient allowance.

6.7 Future prospectus and challenges

The real challenge for the production of meat analogue is to attain same sensory attributes like taste, texture, smell, aroma, etc.). Different food researcher and scientist introduced various techniques such as extrusion and shearing were used to produce desired texture. Another challenge was discussed is in terms of nutritional profile and only 30% protein was reported in meat alternatives. However, Protein Digestibility Corrected Amino Acid Score (PDCAAS) should be focused further for complete replacement of meat analogues with meat products. Not only plant, but other source of meat analogue like microbes (bacteria, yeast, algae) and mushrooms can be used which may fit with the human diet and consumer taste. But maximum attention can be devoted to improve mouth-feel properties meat substitutes which is low in commercially available meat analogue. The large scale production of meat analogue still a challenge until the low sensory attributes of meat substitutes solved and thereby, it will take time to shift the purchasing habits of consumers. In market terms, future of meat analogues is safe and demand is increasing by passing year. For now, meat substitute are viable only for to those vegan diet depended consumers. Presently, the willingness of consuming meat analogue is between low to moderate and availability of low varieties of meat alternates influence its market success, although demand is growing every year. For all protein producers to have better connectivity to consumers there will need to be better investment in systems for efficiency and quality, overlaid with welfare, environmental and health standards.

7. Conclusion

There is a high demand of meat analogue based on plant protein which is growing every year. The development of meat alternative is useful not only to vegan consuming diet but also focus in the environment sustainability and consumer health safety prospective with higher or at least similar benefits as animal meat products. But still the availability and variety of plant based meat analogue is very limited in global market which may be due to several issues. One of the major concerns is how to get the exact texture and flavor in the meat substitutes, and this has resulted in development of the Umami taste. Furthermore, issues related with production of meat analogues like mouth-feel is biggest barrier in the acceptance and replacement of real meat products from the market. Another point is that cultured meat products are not promoted in markets till date due to its complex production procedures and unreported health concerns and their safety issues. The production of meat analogue using plant protein can be considered as a safer and sustainable way to provide the health conscious people an alternate to balance the recommended dietary allowance specially for the proteins without the concerns of the saturate fat intakes. Further, the market trends suggests that there is a huge potential of such innovative food materials in the market which is further expected to grow in the near future as some of the big players of the food processing sector are investing heavily in this sector for the potential improvement in the existing products or the development of the new innovative products for all.

Acknowledgements

The authors are thankful to Lovely Professional University Phagwara, Punjab Agricultural University Ludhiana and Dr. YS. Parmar University of Horticulture and Forestry Solan for providing infrastructure and financial support for the study.

Conflict of interest

The authors declare that they have no conflict of interest.

References

[1] McMichael AJ, Powles JW, Butler CD, and Uauy R. (2007). Food, livestock production, energy, climate change, and health. Lancet, 370, 1253-1263.

[2] Norat, T., Bingham, S., Ferrari, P., Slimani, N., Jenab, M., Mazuir, M., Overvad, K., Olsen, A., Tjønneland, A., Clavel, F. and Boutron-Ruault, M.C., 2005. Meat, fish, and colorectal cancer risk: The European prospective investigation into cancer and nutrition. Journal of the National Cancer Institute, *97*(12), pp.906-916.

[3] Rohrmann, S., Overvad, K., Bueno-de-Mesquita, H.B., Jakobsen, M.U., Egeberg, R., Tjønneland, A., Nailler, L., Boutron-Ruault, M.C., Clavel-Chapelon, F., Krogh, V. and Palli, D., 2013. Meat consumption and mortality-results from the European prospective investigation into cancer and nutrition. BMC Medicine, *11*(1), p.63.

[4] Bouvard, V., Loomis, D., Guyton, K.Z., Grosse, Y., Ghissassi, F.E., Benbrahim-Tallaa, L., Guha, N., Mattock, H., Straif, K. and Corpet, D., 2015. Carcinogenicity of consumption of red and processed meat. The Lancet Oncology, 16(16), pp.1599-1160

[5] IARC (The International Agency for Research on Cancer). 2015. World Health Organization, Q&A on the carcinogenicity of the consumption of red meat and processed meat. *2015-10-29)[2015-12-02]. http://www. cancer. ie/content/qa-carcinogenicityconsumption-red-meat-and-processed-meat# sthash. iz4J5AxV. dpbs*.

[6] Micha, R., Khatibzadeh, S., Shi, P., Andrews, K.G., Engell, R.E. and Mozaffarian, D., Chronic Diseases Expert, G.(2015). Global, regional and national consumption of major food groups in 1990 and 2010: A systematic analysis including 266 country-specific nutrition surveys worldwide. *British Medical Journal*, *5*(9), p.e008705.

[7] Ismail, I., Hwang, Y.H. and Joo, S.T., 2020. Meat analog as future food: A review. Journal of Animal Science and Technology, *62*(2), p.111.

[8] Steinfeld, H., Gerber, P., Wassenaar, T., Castel, V., Rosales, M. and De Haan, C. (2006). In: Livestock's long shadow: Environmental issues and options. Rome, Italy: FAO. FAO 978-92-5-195571-7

[9] Bruinsma, J. (2009). The resource outlook to 2050: By how much do land, water and crop yields need to increase by 2050? Rome, Italy: FAO. In: Expert Meeting on How to feed the World in 2050. http://www.fao.org/wsfs/forum2050.

[10] Kumar, P., Chatli, M. K., Mehta, N., Singh, P., Malav, O. P., &Verma, A. K. (2017a). Meat analogues: Health promising sustainable meat substitutes. Critical Reviews in Food Science and Nutrition, *57*(5), 923-932.

[11] Kumar, P., Chatli, M. K., Mehta, N., Singh, P., Malav, O. P., &Verma, A. K. (2017b). Meat analogues: Health promising sustainable meat substitutes. Critical Reviews in Food Science and Nutrition, *57*(5), 923-932.

[12] Siegrist, M., & Hartmann, C. (2019). Impact of sustainability perception on consumption of organic meat and meat substitutes. Appetite, *132*, 196-202.

[13] Malik, V. S., Li, Y., Tobias, D. K., Pan, A., & Hu, F. B. (2016). Dietary protein intake and risk of type 2 diabetes in US men and women. American Journal of Epidemiology, *183*(8), 715-728.

[14] Mariotti, F. (2019). Animal and plant protein sources and cardiometabolic health. Advances in Nutrition, *10*(Supplement_4), S351-S366.

[15] Tziva, M., Negro, S. O., Kalfagianni, A., &Hekkert, M. P. (2020a). Understanding the protein transition: The rise of plant-based meat substitutes. Environmental Innovation and Societal Transitions, *35*, 217-231.

[16] Tziva, M., Negro, S. O., Kalfagianni, A., &Hekkert, M. P. (2020b). Understanding the protein transition: The rise of plant-based meat substitutes. Environmental Innovation and Societal Transitions, *35*, 217-231.

[17] Zaheer, K., &Humayoun Akhtar, M. (2017). An updated review of dietary isoflavones: Nutrition, processing, bioavailability and impacts on human health. Critical Reviews in Food Science and Nutrition, *57*(6), 1280-1293.

[18] Dukariya, G., Shah, S., Singh, G., & Kumar, A. (2020). Soybean and its products: Nutritional and health benefits. J Nut Sci Heal Diet, *1*(2), 22-29.

[19] Scherf, K. A. (2019). Immunoreactive cereal proteins in wheat allergy, non-celiac gluten/wheat sensitivity (NCGS) and celiac disease. Current Opinion in Food Science, *25*, 35-41.

[20] Albuquerque, T. G., Nunes, M. A., Bessada, S. M., Costa, H. S., & Oliveira, M. B. P. (2020). Biologically active and health promoting food components of nuts, oilseeds, fruits, vegetables, cereals, and legumes. In *Chemical Analysis of Food* (pp. 609-656). Academic Press.

[21] Hashempour-Baltork, F., Khosravi-Darani, K., Hosseini, H., Farshi, P., &Reihani, S. F. S. (2020). Mycoproteins as safe meat substitutes. Journal of Cleaner Production, *253*, 119958.

[22] Souza Filho, P. F., Andersson, D., Ferreira, J. A., &Taherzadeh, M. J. (2019). Mycoprotein: Environmental impact and health aspects. World Journal of Microbiology and Biotechnology, *35*(10), 147.

[23] Ajwalia, R. (2020). Meat alternative gaining importance over traditional meat products: A review. Food and Agriculture Spectrum Journal, *1*(2).

[24] Ibañez, E., Herrero, M., Mendiola, J. A., & Castro-Puyana, M. (2012). Extraction and characterization of bioactive compounds with health benefits from marine resources: macro and micro algae, cyanobacteria, and invertebrates. In *Marine bioactive compounds* (pp. 55-98). Springer, Boston, MA.

[25] Smetana, S., Mathys, A., Knoch, A., & Heinz, V. (2015). Meat alternatives: Life cycle assessment of most known meat substitutes. The International Journal of Life Cycle Assessment, *20*(9), 1254-1267.

[26] Kyriakopoulou, K., Dekkers, B., & van der Goot, A. J. (2019). Plant-based meat analogues. In *Sustainable meat production and processing* (pp. 103-126). Academic Press.

[27] Wang, X., Luo, K., Liu, S., Zeng, M., Adhikari, B., He, Z., & Chen, J. (2018). Textural and rheological properties of soy protein isolate tofu-type emulsion gels: Influence of soybean variety and coagulant type. Food Biophysics, *13*(3), 324-332.

[28] Mani, V., & Ming, L. C. (2017). Tempeh and other fermented soybean products rich in isoflavones. In *Fermented foods in health and disease prevention* (pp. 453-474). Academic Press.

[29] Amin, M. N. G., Kusnadi, J., Hsu, J. L., Doerksen, R. J., & Huang, T. C. (2020). Identification of a novel umami peptide in tempeh (Indonesian fermented soybean) and its binding mechanism to the umami receptor T1R. *Food Chemistry*, 127411.

[30] Alamu, E. O., &Busie, M. D. (2019). Effect of textured soy protein (TSP) inclusion on the sensory characteristics and acceptability of local dishes in Nigeria. Cogent Food & Agriculture, *5*(1), 1671749.

[31] Pintado, T., & Delgado-Pando, G. (2020). Towards more sustainable meat products: Extenders as a way of reducing meat content. Food, *9*(8), 1044.

[32] Anwar, D., &Ghadir, E. C. (2019). Nutritional quality, amino acid profiles, protein digestibility corrected amino acid scores and antioxidant properties of fried tofu and seitan. Food and Environment Safety Journal, *18*(3).

[33] Finnigan, T., Needham, L., & Abbott, C. (2017). Mycoprotein: a healthy new protein with a low environmental impact. In *Sustainable protein sources* (pp. 305-325). Academic Press.

[34] Malav, O. P., Talukder, S., Gokulakrishnan, P., & Chand, S. (2015). Meat analog: A review. Critical Reviews in Food Science and Nutrition, *55*(9), 1241-1245.

[35] Asgar, M. A., Fazilah, A., Huda, N., Bhat, R., & Karim, A. A. (2010). Nonmeat protein alternatives as meat extenders and meat analogs. Comprehensive Reviews in Food Science and Food Safety, *9*(5), 513-529.

[36] Arora, B., Kamal, S., & Sharma, V. P. (2017). Effect of binding agents on quality characteristics of mushroom based sausage analogue. Journal of Food Processing and Preservation, *41*(5), e13134.

[37] Gwiazda, S., Noguchi, A., &Saio, K. (1987). Microstructural studies of texturized vegetable protein products: Effects of oil addition and transformation of raw materials in various sections of a twin screw extruder. Food Structure, *6*(1), 8.

[38] Cheftel, J. C., Kitagawa, M., &Queguiner, C. (1992). New protein texturization processes by extrusion cooking at high moisture levels. Food Reviews International, *8*(2), 235-275.

[39] Pimentel, D., & Pimentel, M. (2003). Sustainability of meat-based and plant-based diets and the environment. The American Journal of Clinical Nutrition, *78*(3), 660S–663S.

[40] Key, T. J., Davey, G. K., & Appleby, P. N. (1999). Health benefits of a vegetarian diet. Proceedings of the Nutrition Society, *58*(2), 271-275.

[41] Larsson, S. C., &Wolk, A. (2006). Meat consumption and risk of colorectal cancer: A meta-analysis of prospective studies. International Journal of Cancer, *119*(11), 2657-2664.

[42] Cho, M. J. (2010). Soy protein functionality and food bar texture. In *Chemistry, texture, and flavor of soy* (pp. 293-319). American Chemical Society.

Chapter 7

Breeding Vegetables for Nutritional Security

Abstract

The most dominant vegetables in the global food economy are tomato, cucurbits, (pumpkin, squash, cucumber and gherkin), allium (onion, shallot, garlic) and chili. These vegetables are consumed in nearly all countries although with much variation in shape, size, color and taste, while the marketing of global vegetables accounts for significant revenue streams, traditional vegetables often have superior nutritional properties. Biodiversity is considered essential for food security and nutrition and can contribute to the achievement through improved dietary choices and positive health impacts Through conventional breeding approach, it is possible to develop new vegetable varieties or integrate the favorable genes for neutraceuticals, bioactive compounds and edible color into cultivated varieties. Advances in molecular biology and recombinant technology have paved the way for enhancing the pace of special trait variety development using marker assisted breeding and designing new vegetable crop plants following transgenic approach.

Keywords: biodiversity, neutraceuticals, conventional breeding, MAS, transgeneics

1. Introduction

Vegetables are increasingly recognized as an essential source for food and nutrition security. Vegetable production provides a promising economic opportunity for reducing rural poverty and unemployment in developing countries and is a key component of farm diversification strategies. Vegetables are mankind's most affordable source of vitamins and minerals needed for good health. Today, neither the economic nor nutritional power of vegetables has not been sufficiently realized. To tap the economic power of vegetables, Governments will need to increase their investment in farm productivity (including improved varieties, alternatives to chemical pesticides, and the use of protected cultivation), good postharvest management, food safety, and market access. To tap the nutritional power of vegetables consumers need to know how vegetable consumption must therefore be nurtured through a combination of supply side interventions and communication emphasizing the importance of eating vegetables, governments and donors will need to give vegetables for good nutrition and health, to fully tap the economic and nutritional power of vegetables, much greater priority than they currently receive. Now is the time to prioritize investments in vegetables, providing increased economic opportunities for smallholder farmers and providing healthy diets for all.

Fruits and vegetables are essential sources for the micronutrients needed for healthier diets. Potassium in vegetables helps to maintain healthy blood pressure, their dietary fiber content reduces blood cholesterol levels and may lower the risk of heart disease, folate reduces the risks of birth defects, and vitamin A keeps eyes and skin healthy, while vitamin C not only keeps teeth and gums healthy but also aids in iron absorption. Recognizing the important nutritional benefits of fruits and vegetables, the World Health Organization (WHO) recommends a minimum intake of 400 g per day to prevent chronic diseases (especially heart diseases, cancers and diabetes) and supply needed micronutrients (especially calcium, iron, iodine, vitamin A and Zinc). However, consumers today even those with higher incomes, are believed to be missing this target. More attention in filling this dietary gap and enabling consumers to take the nutritional power of vegetables is required.

2. Vegetables in global food economy

The most dominant vegetables in the global food economy are tomato, cucurbits, (pumpkin, squash, cucumber and gherkin), allium (onion, shallot, garlic) and chili. These vegetables are consumed in nearly all countries although with much variation in shape, size, color and taste, while the marketing of global vegetables accounts for significant revenue stream, traditional vegetables often have superior nutritional properties. For instance, 100 g of leaves of amaranth or vegetable cowpea can provide over 100% of the vitamin A needs of pregnant women. Globally, a one per cent increase in per capita income in developing countries is associated with a 0.5% increase in per capita vegetables availability. It follows that the bulk of the global supply of fruit and vegetables (77% of total value) is produced in populous middleincome countries. China accounts for 45% of the global value of vegetable production. India comes second, accounting for eight per cent of global vegetable production. Vegetables play a major role in world agriculture by providing food and offering nutritional and economic security.

3. Health promoting nutrient compounds in vegetables

Macro carotene, nutraceuticals, phyto-chemicals are bioactive compounds which are either plant pigments (lycopene, ß carotene, anthoyanin, lutein, capsanthin, zeaxanthin *etc.*) or secondary metabolites (Flavonoids, isothiocyanates, glucosinolates *etc.*) found in most of the vegetables which provide color, flavor and texture and protect the plants

Antioxidants: Phyto-chemicals with antioxidant activity are carotenoids (a-carotene, lycopene, zeaxanthin, cryptoxanthin, lutein) flavonoids and polphenols present in vegetables.

Flavanoids: Vegetables such as kale, spinach, Brussel's sprouts, sprouting broccoli, beets, red bell-pepper, onion, corn, eggplant, cauliflower and cucumber are also rich source of flavonoids hence, have potent antioxidant activity.

Plant pigments: Plant pigments are edible colors found in tissue of plants which includes anthocyanins, betalains, carotenoids and chlorophylls.

Edible colors: Edible colors are natural pigments found in tissue of plants. These colors are the chemical compounds produced by several biochemical pathway and gives colors to the food. These colors are changed according the growth stage of the plant parts or vegetable product. These include anthocyanins, betalains, carotenoids, chlorophylls. These pigments play important ecological and metabolic functions in the plants [1] and are more frequently exploited as the source of nutraceuticals to address a number of human ailments.

Betalains: Betalains have been widely used as natural colorants for many centuries but their attractiveness for use as colorants of foods (or drugs and cosmetics) has increased recently due to their reportedly high anti-oxidative free radical scavenging activities, and concerns about the use of various synthetic alternatives.

Chlorophylls: Chlorophyll is the most important plant pigment which has the potential to act as a chemo-preventive compound in humans. Chlorophylls, in contrast, are typically consumed in much higher doses in a diet that incorporates green and leafy vegetables [2]. The anti- mutagenic properties of chlorophylls have neen demonstrated in various assays, and clearly, intake of chlorophyll has potential to act as a chemopreventive compound in humans.

Anthocyanins: Anthocyanins are natural pigments belonging to the flavonoid family. They are responsible for the blue, purple, red and orange color of many fruits and vegetables. Anthocyanins are capable of acting on different cells involved in the development of atherosclerosis, one of the leading causesa to cardiovascular dysfunction. Anthocyanins and the aglycone cyaniding were found to inhibit cycloxygenase enzymes, which can be one market for the initiation stage of carcinogenesis. On one hand they can interfere with glucose absorption and on the other hand they may have a protective effect on pancreatic cells. The most extensively documented phytomedicinal role of anthocyanin pigments is in improving eyesight including night vision.

Carotenoids: Carotenoids are the second most abundant pigments in nature, and consist of more than 700 members [3]. Carotenoids play an important role in plant reproduction, through their role in attracting pollinators and in seed dispersal, and are essential components of human's diets. Carotenoids provide protection to vision and eye function, and against macular degeneration and cataracts. Carotenoids are credited with biological promotion of immune system response. Carotenoids are associated with inhinition of several types of cancers including cervical, esophageal, pancreatic, lung, prostate, colorectal and stomach.

4. Antinutrient factors

Plants produce many defense strategies to protect themselves from predators and many of these, such as resveratrol and glucosinate, which are primarily pathogen-protective chemicals also have demonstrated beneficial effects for human health. Many, however, have the opposite effect. For example, phytate, a plant phosphate storage compound, is an antinutrient as it strongly chelates iron, calcium, zinc and other divalent mineral ions making them unavailable for uptake. Different antinutrient compounds (phytates, oxalates, trypisn inhibitors, lectins etc.), food allergens (albumins, globluins etc.) and toxins (glycoalkaloids, cynogenic glucosides, phyto-hemaglutinins) in crop plants need to be reduced to enhance nutrient potential of the vegetables.

5. Breeding nutrient rich varieties

Vegetables are valued for their extrinsic and intrinsic quality traits. Diet rich in vegetables provide micronutrients and health promoting phytochemicals that alleviate malnutrition. The beneficial health effects are mainly attributed to diverse antioxidant compounds such as vitamins, carotenoids, phenolics. Alkaloids, nitrogen containing compounds, organouslphur compounds etc. Although, chief long term breeding objective will continue to be increasing yield to meet the food requirement of ever increasing population, in order to ensure health security to our countrymen and multipurpose utility of the varieties for fresh market and industry suitability, it is imperative that nutraceutical, edible color and bioactive compound rich vegetable varieties are bred ensuring high remuneration to farmers. Quality in vegetables is a complex character influenced by both genetical and environmental factors. Breeding for quality has been unsystematic and often empirical but significant progress has been made in several vegetable crops. Conventional breeding in conjunction with molecular biology has bright prospects of developing vegetable varieties high in neutraceuticals, edible colors and bioacticve compounds suitable for fresh market as well as developing functional fusions food industry.

Conventional breeding uses inherent properties of the crop, having far reaching impact on communities and has fewer regulatory constraints compared to genetically modified varieties. Breeding efforts targeting improved micronutrient content and composition began in the 1940s and 1950's, with research describing the inheritance and development of tomato breeding stocks high in pro-vitamin A carotenoids and vitamin C. Similar research leading to the development of darker orange and consequently high pro-vitamin A, carrots began in the 1960s. Since then genetic improvement to increase levels of specific micronutrient has been pursued primarily in several vegetables.

A significant genetic component of iron and zinc content of edible plant parts has been noted, but parallel investigations for calcium are not widely reported for many plant species and even less is known about magnesium. As progress is made in breeding for crop yield, mineral content usually is reduced. Furthermore, breeding for improved mineral use efficiency usually does not alter mineral content of edible plant parts. Success in breeding for higher mineral content must consider not only mineral concentration but also organic components in plants that can be abundant and either reduce (phytate, phenolic compounds) or increase (vitamin C) bioavailability. Recent studies have exhibited a broad range of calcium, iron, and zinc content across a range of Andean potato cultivars [4].

5.1 Genetic resources

Biodiversity is considered essential for food security and nutrition and can contribute to the achievement through improved dietary choices and positive health impacts. However, it is seldom included in nutrition programmes and interventions. Dietary diversity depends not only on a diversity of crops but also on diversity within crops. There is an increasing body of evidence of wide variation in nutrient contents within species, but data are lacking on nutrient composition and dietary intake for many underutilized species as well as for cultivars within species. Such information is needed both to enhance use of more nutritious cultivars in diets and to make them available for use in breeding programme aimed at increasing the nutrient content of more commonly used varieties for the same species, eliminating the need for transgenic modifications.

Genetic resources are the foundation block that are essentially required for evolving improved crop varieties when the breeder aim at adding more desirable traits to an otherwise acceptable varieties. This necessitates availability of the desired variability to the breeder within the land races, putative ancestral form, primitive cultivars and obsolete cultivars, heirloom cultivars of these crops or its wild forms and other elated species constituting primary, secondary and tertiary gene pools. The utilization of plant genetic resources to enhance the chemical composition of horticultural crops through biotechnology or conventional breeding has led to the development of varieties with enhanced levels of micronutrients, such as enhanced beta-carotene sweet potatoes, potato, carrot. Pavithra *et al.* [5] found tomato lines rich in Zn content. The elite germplasm line with high Zn content may be used to prospect candidate gene for improving nutritional value.

5.2 Breeding for nutraceutical bioactive compounds

Nutraceutical bioactive and edible colors are natural compounds which are regulated by several biochemical pathways and controlled by genetical and environmental factors. From early times people knowingly or unknowingly selected several vegetable crops for their food purpose. There are many or cultivated vegetables which are rich in these beneficial compounds. The biochemical pathway and synthesis of these compounds are controlled by one or many genes which are scattered in the available or unknown germplasm of particular vegetable crop. India is endowed with diverse agroclimatic regions ranging from tropical to temperate making it possible to grow all kinds of vegetable crops in one or the other corners of the country. Besides, there is plenty of diversity in different vegetable crops which can be exploited for development of special varieties. Through convention breeding, it is possible to develop new vegetable varieties or integrate the favorable genes for nutraceuticals, bioactive compounds and edible color into cultivated varieties, advances in molecular biology and recombinant technology have paved the way for enhancing the pace of special trait variety development using marker assisted breeding and designing new vegetable crop plants following transgenic approach.

5.3 Breeding techniques

Breeding method in any crops depends upon the breeding system and genetic architecture resulting from natural selection as well as human selection during the course of cultivation. The genetic architecture or the pattern of inheritance of characters is another important consideration while determining the most appropriate breeding procedure applicable to any particular crops. The choice of breeding method would be largely guided by nature of gene action and relative magnitude of additive genetic variance, dominance variance and epistasis in a breeding population. The efficient breeding procedure should be effective in manipulation and selection of favorable gene combination, additive genetic variance, exploitation of dominance variance and achieving close relationship between expected genetic gain and realized progress from selection. Development of F_1 hybrid is very suitable for enhancing nutraceuticals and edible colors. The beta-carotene content in muskmelon has increased manifold in F_1 hybrid [6].

5.4 Advanced breeding techniques

5.4.1 Mutation breeding

In a simple way, mutation is a random or directed change in the structure of DNA or the chromosome which often result in a visible or detectable chance in specific character or trait. In self-pollinated crops, it is well known whereas in cross pollinated crops its application is more difficult and identification of the origin of the desirable genotypes is difficult. Sapir et al. [7] reported in tomato that *high pigment 1* (*hp-1*) mutation known to increase flavanoids content in fruits.

5.4.2 Polyploidy breeding

Polyploid can be induced due to aberration in cell division. This may occur both in the mitosis as well as in meiosis. This method can be used successfully in vegetable breeding as a means of enhancing nutraceuticals and colors in vegetables.

Tetraploids in radish, pumpkin, and watermelon are highly productive and have improved quality. Zhang *et al*. [8] developed tetraploid muskmelon which is rich in vitamin C which is higher than those in the diploid fruit.

5.4.3 Haploidy breeding

The development of haploids in a number of plant species is now recognized as the most rapid route to the achievement of homozygosity and production of pure lines. Currently, little breeding effort is going on for improvement of *brassica* for nutraceutical species and as indicated, there are very few successful double haploid protocols.

5.5 Biotechnological approaches

5.5.1 Molecular markers and marker assisted selection (MAS)

Molecular markers such as RAPD, ISSR, SSR, SCAR, CAPS are used to study linkage with gene(s) responsible for high neutraceuticals, bioactive compounds and edible colors using mapping population. Of late, use of SNP marker is becoming more common. In Marker Assited Selection, a marker (morphological, biochemical or DNA) is used for indirect selection of a trait of interest. The mapping populations such as Near Isogenic Lines (NILs), Recombinant Inbred Lines (RILs) are used to identify the molecular markers linked to genes of interest. Ripley and Roslinsky [9] identified an ISSR Marker for 2 propenyl glucosinolate content in *Brassica*.

5.5.2 Quantitative trait loci (QTL) analysis

QTL analysis is the study of the alleles that occur in a locus and the phenotypes. Most traits of interest are governed by more than one gene, defining and studying the entire locus of genes related to a trait gives hope of understanding the effect genotype of an individual. The advent of molecular maps and the derived quantitative trait locus (QTL) mapping technology has provided strong evidence that despite the inferior phenotype, exotic germplasm is likely to contain QTLs that can increase the quality of elite breeding lines. Bin 3-C has previously been described as harboring a single gene mutation *r* yellow flesh in tomato [10].

5.5.3 Advanced backcross QTL analysis

The AB-QTL strategy has so for been tested in tomato and pepper. The most extensive experiments have been conducted in tomato, where populations involving crosses with five wild *Lycopersicon* species have been genotyped and field tested in a number of locations around the world for numerous traits important for the tomato processing industry. Through the application of marker and phenotypic analysis of segregating generations of the cultivated tomato and wild *Lycopersicon* species, QTLs for improved fruit color had been revealed. QTLs that improve fruit color originating from red-fruited (*S.pimpinellifolium*) and green fruited (S.*habrochaites, S parviflorum*) wild relatives had been detected in segregating populations of crosses of these species and the cultivated tomato. Quantitative trait loci associated with carotenoids and tomato fruit color using introgression populations of *S.pennellii*, *S peruvianum* and *S. habrochaites* have been described by [11].

5.5.4 Introgression line (IL) libraries

IL libraries contain homogenous genetic backgrounds, only differing from one another by the introgressed donor segment. A tomato introgression line population that combines single chromosomal segments introgrossed from the wild, green fruited species *Solanum pennellii* in the background of the domesticated tomato, *Solanum lycopersicum* was used to identify QTL for nutritional and antioxidant contents. Liu *et al.* [12] applied the candidate gene approach to link sequences that have known functional roles in carotenoid biosynthesis to QTLs that are responsible for the variation of the tomato red fruit color.

Marker assisted backcross breeding has been used successfully to incorporate genes or QTL for both qualitative and quantitative traits in a number of crop species especially tomato, cucumber, potato, in some cases leading to the development of improved cultivars. Of late Indian cauliflowers are being introgressed with semi-dominant mutant *Or* gene to enhance their betacarotene content in an attempt to tackle malnutrition problem by making diverse beta- carotene rich food available to consumers.

Interspecific crosses with wild species transferred the ability to produce small quantities of anthocyanins into the peel of cultivated tomatoes. For example, the dominant gene Anthocyanin fruit (*Aft*), which induces limited pigmentation upon stimulation by high light intensity, was introgressed into domesticated tomato plants by an interspecific cross with *S. chilense* and the gene Aubergine (*Abg*) from *Solanum lycopersicoides,* Furthermore, the recessive gene atroviolacea (*atv*), derived from the interspecific cross with *Solanum cheesmaniae* that stimulate strong anthocyanin pigmentation in the entire plant, particularly in vegetative tissues. Fruits with either *Aft* and *atv* alleles or *abg* and *atv* alleles have been obtained with higher production of anthocyanins in the peel, ranging in total amount from 1 to 4 mg/g fresh weight of peel. Anthocyanins were found in the skin and flesh of certain cultivars of potato. Total anthocyanin concentrations in Andean potatoes ranged from 14 to 16, 330 µg/g DW [4]. Usually, cultivars high in anthocyanins are low in carotenoids and *vice versa*. The fruit color of red chili is genetically determined by three loci *y*, *cl*, and *c2* Recently the gene for capsanthin-capsorubin synthase (CCs) has been considered as candidate gene for the *y* locus. The relationship between the phytoene synthase and carotenoid content in chili was tested with interval mapping using QTL analysis revealed that they were detected only at the *PSY* locus.

Singh *et al*. [13] observed enormous diversity in pigmentation of European and Asiatic carrot.The Asiatic types are mostly yellow and purple. The Asiatic type collection Local Rewari Black and Local Jaipur Black have higher anthocyanin content. Few molecular markers linked to major genes or QTL have been developed for carotene [14] and the Y_2 gene and the *Rs* sugar type gene. To date, seven monogenic traits have been mapped for carrot: *yel, cola, Rs, Mj-1,* Y,Y_2 and P_1.QTL have been mapped for carrot total carotenoids and five component carotenoids: phytoene, x-carotene, ß-carotene, zeta-carotene, and lycopene [14] and the majority of the structural genes of the carotenoid pathway are now placed into this map. Anthocyanin accumulation in the carrot phloem is conditioned by the P_1 locus, with purple (P_1) dominant to non purple (p_1). From the inheritance studies of Eastern carrot germplasm, it is concluded that the P_1 and Y_2 loci are unlinked.

The common cucumbers always develop white fruit with lower carotenoid, 22 48 µg/100 g fresh weight. While Xishuangbanna gourd (*C sativus var.xishuangbannanesis*) develops orange fruit rich in carotenoid, approximately 700 µg/100 g flesh weight, which makes this germplasm attractive to plant improvement programme. QTL associated with orange color fruit flesh showed two genetic linkage maps with the markers of RAPD, SCAR, SSR, EST, SNP, AFLP and SSAP, which defined a common collinear region containing four molecular markers on linkage group LG6 inMAP1 and LG 3 in Map 2.

SCAR markers linked to the *Or* gene were identified based on random amplified polymorphic DNA (RAPID) and amplified fragment length polymorphism (ALPH) by performing a bulked segregant analysis (BSA) using a double haploid (DH) population derived from the F_1 cross between 91 and 112 (white head leaves) and T12–19 (orange head leaves) *via* microspore culture. On the basis of linkage analysis, the *Or* gene was mapped in a region conversing a total interval of 4.6 cM between two SSR markers derived from BAC clones AC172873 and AC189246 at the end of linkage group 9, which matches with chromosome I of A genome in Chinese cabbage. A genetic map of the 'or' locus was constructed by using five SSR markers and two morphological markers. Three SSR markers were tightly linked to *or* and two of them, *sau* (C) 586 and *syau* 19, were located on the same side at distances of 1,6 and 1,3 cM, respectively. The other marker, *syau* 15, was located on the other side at a distance of 3.3 cM. Cervantes-Flores et al. [15] have recently reported QTL for dry matter, starch content and ß-carotene content, opening up the possibility of genetic manipulation and further enhancement of sweet potato. Ripley, V.L. and Roslinsky, V [9] identified ISSR marker for 2- propenly glucosinolate content in Brassica. Efforts are also being made to use the genetic and molecular approaches for increasing the leaves of tocopherols in potato tubers through metabolic engineering tools and techniques.

6. Transgenic approach

Three genes, encoding phytoene synthase (*CrtB*), phytonene desaturase (*Crt1*) and lycopene beta – cylase (*CrtY)* from *Erwinia* have been introduced in potato to produce beta carotene. Romer *et al*. [16] developed transgenic tomato to enhance the carotenoid content with the bacterial carotenoid gene (*crtl*) encoding the enzyme phytoene desaturase, which converts phytoene into lycopene. Lu et al. [17] suggested that transgenic cauliflower with *Or* transgene is associated with a cellular process that triggers the differentiation of proplastids into chromoplasts for carotenoid accumulation and *Or* can be used as a novel genetic tool to induce carotenoid accumulation in a major staple food crops.

One of the most obvious benefits of enhancing carotenoid levels is the increase in pigmentation, which can lead to more deeply colored vegetables that are often preferred by consumers. Thus, increasing levels of carotenoid is doubly beneficial, both in terms of nutrition and esthetics. There are a range of other approaches to enhance the carotenoid levels in potatoes and other root vegetables. Diretto *et al.* [18] have silenced the first step in the ß-epsilon branch of carotenoid biosynthesis, lycopene epsilon cyclase (LCY-e) in potato which is low in carotenoids. This antisense tuber-specific silencing of the gene results in significant increases in carotenoid levels, with up to 14-fold more ß-carotene.

Enhancing anthocyanins: Potato does not normally produce anthocyanin, but germplasm expressing anthocyanin pigment has been developed and is attracting interest from consumers. One of these genes which encode a novel single -domain MYB transcription factor has the potential to influence anthocyanin pigment production in potato. The resulting purple potato might offer both novelty and health functionality to consumers, who can also benefit from native Andean potatoes that do not always show desired tuber shapes for both table and processing industry.

Folates rich tomato: Diaz de la Garza et al. [19] developed transgenic tomatoes by engineering fruit specific over expression of GTP cyclohydrolase I that catalyzes the first step of pteridine synthesis, and amino deoxychorismate synthase that catalyzes the first step of PABA synthesis. Vine ripened fruits contained on average 25 fold more folate than controls by combinig PABA and pteridine overproduction traits through crossbreeding of transgenic tomato plants.

The extent to which vegetable brassicas protect against cancer probably depend on genotype of the consumer, in particular the allele present at the GSTM 1 locus. This gene codes for the enzyme glutathione transferase, which catalyzes the conjugation of glutathione with isothiocyanates. Approximately, 50% of humans carry a deletion on the GSTM1 gene which reduces their ability to conjugate, process and excrete isothiocyanates. Individuals with two null alleles for GSTM1 might gain less protection from these cultivars of vegetable. The most commonly consumed *Brassica* vegetable in Asia is *Brassica rapa*. *B. rapa* contains different isothiocyanates to *B. oleracea* and recent evidence suggests that individuals who are null for GSTM1 can gain a protective benefit from *B. rapa* [20].

A 10 fold increase in the level of 4-methylsulphinylbutyl glucosinolate was obtained by crossing broccoli cultivars with selected wild taxa of the *Brassica oleracea* (chromosome number, n = 9) complex. Tissue from these hybrids exhibited 100 fold increase in the ability to induce quinone reductase in Hepa1cIc7 cells over broccoli cutlivars, due to an increase in 4 – methylsulphinylbutyl glucosinolate content.

Vegetables of the *Allium* genus such as onion, garlic, leek and chive are among the oldest crops associated with health-related properties. Three sets of transgenic onion plants containing antisense alliinase gene constructs (a CaMV 35S-driven antisense root alliinase gene, a CaMV 35S-driven antisense bulb alliinase, and a bulb alliinase promoter-driven antisense bulb alliinase) have been recently produced [21]. Transgenic hybrid onion seed from these transgenic lines has been developed by crossing a nontransgeneic open- pollinated parental line with a transgenic parental plant carrying a single transgene in the hemizygous state.

Miraculin rich vegetables: For reduction of bitterness in lettuce, the gene for sweetness and taste modifying protein miraculin, from the pulp of berries of West African shrub *Richadella dulcifica* was cloned [22]. This gene, with the CaMV 35S promoter, was introduced into the lettue cultivar "Kaiser" using *A. tumefaciens* GV2260. Expression of this gene in transgenic plants led to the accumulation of significant concentrations of the sweet enhancing protein.

Protein rich potato: The genetically modified potato developed at CPRI in collaboration with NIPGAR "Protato" contains 60% enhanced protein content. This has been achieved by introducing *AmA1* gene (*Amaranth Albumin 1*) from edible amaranth plant into seven commercial varieties of potatoes. The GM potato plants were tested in India and the results demonstrated greater harvest and moderate increase in tuber yield. Safety evaluation indicated that the transgenic potatoes are suitable for commercial cultivation and have no negative effects on animal health. In addition, the concentration of several essential amino acids increased significantly in transgenic tubers which are otherwise limited in potato. This resulted in a significant increase in yield and enhanced nutrition. The *AmA1* gene has been reported to have potential for the nutritional improvement of other food crops as well [23].

7. Conclusion

Vegetables are nutritional powerhouses, key sources of micronutrients needed for good health. They add diversity, flavor and nutritional quality to diets.

A strengthened focus on vegetables may be the most direct affordable way to deliver better nutrition for all. Intensified vegetable production has the potential to generate more income and employment than other segments of the agricultural economy, making vegetable an important element of any agricultural growth strategy. Today neither the economic nor the nutritional power of vegetables is sufficiently realized. With a growing understanding of the linkages between dietary quality and health, policy makers must also be prepared to support additional interventions to promote vegetable consumption. Breeding for improved taste, convenience, nutritive value and consumer appeal has already contributed in increase per capita vegetable consumption with the development of products such as baby carrots, yellow and orange peppers, cherry and pear tomatoes, seedless watermelons and lettuces with different with different color, texture and flavor. Therefore conventional breeding in conjunction with molecular biology has bright prospects of developing high yielding vegetable varieties with high neutraceutcials and bio active compounds suitable to offer nutritional security.

References

[1] Grotewold, W .(2006). The genetics and biochemistry of floral pigments. Annu. Rev., Plant Biol. 57:761-780

[2] Chakraborty, S., Chakrabortya, N., Agrawala, L., Ghosha, S.and Narula, K. 2010. Next generation protein-rich potato expressing the seed protein gene *AmA1* is a result of proteome rebalancing in transgenic tuber. Proceedings of the National Academy of Sciences of the United States of America 107:17533-17538.

[3] Britton, G. (1998). Overview of carotenoid biosynthesis. In: *Carotenoids,* Birkhauser, Basel, Switzerland, pp. 13-147.

[4] Andre, C.M., Ghislain, M., Bertin, P., Qufir, M., Del Rosario Herrera, M., Hcffmann, L., Hausman, J.F., Larondelle, Y. and Evers, D. 2007. Andeen potato cultivars (*Solanum tuberosum* L.) as a source of antioxidant and mineral micronutrients. Journal of Agricultural and Food Chemistry 55:366-378

[5] Pavithra, G.J., Mahesh, S. and Shankar, A.G 2014. Assessment of zinc variability in tomato germplasm lines: An option for biofortification by introgression breeding Agrotechnology, 2:229

[6] Moon, S.S., Verma, V.K. and Munshi, A.D. (2002). Gene action of quality traits in muskmelon *(Cucumis melo L.). Vegetable* Science, 29 (2) 134-136

[7] Sapir, M, Sharmir, M.O., Ovadia, R, Reuveni, M, Evenor, D, Tadmor, Y, Nahon, S. Shlomo, Molecular spects of *Anthocyanin fruit* Tomato in Relation to *high pigment* -1 99 (3) 292-303.

[8] Zhang, W, Hao, H, Ma, L, Zhoa, C and Yu, X (2010). Tetraploid muskmelon improves fruit quality. *Scientia Horticulturae*, 125 (3) 396-400.

[9] Ripley, V.L. and Roslinsky, V.(2005) Identification of an ISSR marker for 2 propenyl glucosinolate content in *Brassica juncea* L. and Vonversion to a SCAR marker. Molecular Breeding 161: 57-66

[10] Fray, R. G, and Grierson, D. (1993). *Molecular genetics of tomato fruit ripening. Trends in Genetics,* (438-443).

[11] Berancchi, D, Beck-Bunn, T., Eshed, Y., Lopez, J., Petiard, V., Uhlig, J., Zamir, D. and Tanksley, S. (1998). Identification of QTLs for traits of agronomic importance from *Lycopersicon hirsutun.* Theoretical and Applied Genetics, 97: 381-397.

[12] Liu, Y.S., Gur, A., Ribeb, A., Causes, M., Damidaux, R., Buret, M., Hirschberg, J. and Zamir, D (2003) There is more tomato fruit colour than candidate carotenoid genes. Plant Biotechnology Journal, 1 (3): 131-240.

[13] Singh, J., Singh, B, and Kalloo, G. 2002 Root morphology and carotene content in Asiatic and European carrots (Daucus carota var. sativa) Indian The Journal of Agricultural Science 72:225-227.

[14] Santos, C.A.F. and Simon, P.W 2002. QTL analyses reveal clustered loci for accumulation of major provitamin a carotenes and lycopene in carrot roots. Mol.Gen.Genomics , 268:122-129

[15] Cervantes-Flores, J.C., Sosinski, B., Pecota, K.V., Mwanga, R.O.M., Catignani, G.L., Truong, V.D., Watkins, R.H., Ulmer, M.R., Yencho, G.C. 2010. Identification of quantitative trait loci for drymatter, starch, and beta carotene content in sweet potato. Molecular Breeding doi:10.1007/s11032-010-9474-5

[16] Romer, S., Fraser, P.D., Kiano, J.W., Shiption, C. A., Misawa, N., Schuch, W., Bramley, P.M. (2000) Elevation of

the pro-vitamin a content of transgenic tomato plants Nature Biotechnology, 18: 666-669.

[17] Lu, S, Eck, J.V., Zhou, X., Lopez, A.B, Halloran, D.M., Cosman, K.M., Conlin, B.J. Gene encodes a Dnaj cysteine-rich domain-containing protein that mediates high levels of ß-carotene Acumulation. The Plant Cell, 18: 394-3605

[18] Diretto, G., Tavazza, R, and Giuliano, G 2006. Metabolic engineering of potato tuber carotenoids through tuber-specific silencing of lycopene epsilon cyclase. BMC Plant Biol. Dol: 10.1186/1471-2229:6-13.

[19] Diaz de la Garza, R., Quinlivan, E.P., Klaus, S.M., Basset, G.J. and Gregory, F. F, 2004. Folate biofortification in tomatoes by engineering the pteridine branch of folate synthesis. Proceedings of the National Academy of Sciences of the United States of America 101: 13720-13725

[20] Gasper, A.V. 2005. Glutathione-S transferase M1polymorphism and metabolism of sulforaphane from standard and high –gkycisubikate vricciku. Am.J.Clinical Nutrition 82:1283-1291

[21] Eady, C. C Kamoi. T and Kato. M. 2008. Silencing onion lachrymatory factor synthase causes a significant change in the sulfur secondary metabolitie profile. Plant Physiology 147:296-2106.

[22] Sun, H.J., Cui, M.L., Ma, B. and Ezura, H. 2006. Functional expression of the taste modifying protein, Miraculin, in transgenic Lettuce, FEBS letters, 580:620-626.

[23] Delgado-Vargas, F and Paredes-Lopez O, eds. 2003 Natural colorants for Food and Nutraceutical Uses. CRC Press, Florida, USA.

Index